FIELD-ION MICROSCOPY

FIELD-ION MICROSCOPY

Edited by **John J. Hren**

Department of Metallurgical and Materials Engineering
University of Florida, Gainesville, Florida

and **S. Ranganathan**

Inorganic Materials Research Division
Lawrence Radiation Laboratory
University of California, Berkeley, California

Based Upon a Short Lecture Course
Presented March 14-22, 1966
at the University of Florida, Gainesville, Florida

ℚ Springer Science+Business Media, LLC 1968

ISBN 978-1-4899-6241-6   ISBN 978-1-4899-6513-4 (eBook)
DOI 10.1007/978-1-4899-6513-4

Library of Congress Catalog Card Number 68-14853

© 1968 Springer Science+Business Media New York
Originally published by Plenum Press in 1968.
Softcover reprint of the hardcover 1st edition 1968

Dedicated to

Professor Erwin W. Müller

by his coauthors for his overwhelming contributions to
the theory and practice of field-ion microscopy.

D. G. Brandon	A. J. W. Moore
J. J. Hren	B. Ralph
A. J. Melmed	S. Ranganathan
M. J. Southon	

PREFACE

The short course on Field-Ion Microscopy held at the University of Florida, Gainesville, Fla., Mar. 14–22, 1966, whose lectures comprise this volume, was intended to be a means of assembling and making accessible the information essential to a research group just taking up field-ion microscopy. As a consequence, the present chapters do not read like the usual proceedings of a symposium but rather somewhat like a graduate-level textbook. Insofar as is possible when there are eight authors, this is exactly what is intended. Not all of the presently known applications of the technique are treated with equal thoroughness, but we hope none has been neglected. The closely related subject of field-electron emission has been given only a cursory treatment here, since other comprehensive treatments on this subject are available.

Although it may be unusual, the other seven authors would like to dedicate this book to their coauthor, Professor E. W. Müller. Inventor of field-electron-emission microscopy and field-ion microscopy and the most important contributor to both fields for many years, Professor Müller still possesses the eagerness, imagination, and drive of a fledgling Ph.D. To observe this personally, one need only attend the annual Field-Emission Symposium. We salute this remarkable scientist.

CONTENTS

Chapter 4 Gas Impact, Field Etch-   D. G. Brandon
ing, and Field Deformation   Battelle Memorial Institute

Chapter 5 Some Geometrical
Aspects of Surfaces Related to   A. J. W. Moore
Field-Ion Microscopy   C.S.I.R.O.

Chapter 9 Field-Ion Microscope   S. Ranganathan
Studies of Interfaces   Lawrence Radiation Laboratory

Chapter 10   Experimental Studies   B. Ralph
of Alloys with Field-Ion Micro-   University of Cambridge
scope

Chapter 11   Field-Ion Microscope   B. Ralph
Studies of Radiation Damage   University of Cambridge

Chapter 12 Field-Ion Microscopy
of Whiskers and Thin Films and
Applications (Real and Imagined)
to Mass Spectrometry and Bio-   A. J. Melmed
logical Molecule Imaging   National Bureau of Standards

APPENDICES

Appendix A The First Fifteen Years of Field-Ion Microscopy
—A Bibliography

Appendix B Lattice Geometry

Appendix C Angles Between Crystallographic Planes in Crystals of

Appendix D Diagrams of Standard and Stereographic Projections

Chapter 1

THE THEORETICAL AND TECHNICAL DEVELOPMENT OF FIELD-ION MICROSCOPY

E. W. Müller†

1.1 Early Historical Development

A field-ion microscope is the most powerful microscopic device known today. It is the only instrument that can show directly the atomic structure of a specimen and the atomic lattice defects. But, for reasons that might lie in the difficulty of operation of the first instruments, perhaps the unorthodoxy of the principles involved, and a justified lack of commercial interest, it took a long time to be developed. When in the spring days of quantum mechanics Gamow[1] (1928) explained the radioactive alpha decay as a tunneling effect, field-electron emission from metals was soon recognized by Fowler and Nordheim[2] as another example of barrier penetration and simultaneously Oppenheimer[3] suggested that the effect of field ionization of free atoms could occur when an electron would tunnel out in the presence of an electric field. While the first two effects commanded considerable interest, field ionization from the ground state of an atom was experimentally inaccessible because of the magnitude of the fields required. Handling large fields became a possibility with the introduction of the field-emission microscope in 1936.[4] With the discovery of field desorption[5] from a positive-point electrode the field range beyond 100 MV/cm, in which all effects of interest to us are taking place, was entered for the first time. The realization that the resolution limit of the field-electron microscope[6] is determined by the tangential velocity of the emitted electrons and, to a lesser extent, by their de Broglie wavelength, which cannot be controlled under the prevailing conditions, led in 1951 to successful imaging of the emitter surface with positive ions rather than electrons.[7] Atomic resolution was thus achieved for the first time. The imaging ions were first thought to originate from an intermediate adsorbed

† Research Professor of Physics, The Pennsylvania State University, University Park, Pa.

1

state, although true field ionization of hydrogen at higher fields in free space was clearly recognized to occur in these early experiments.

Details of the further development of field-ion microscopy since its inception will be more profitably discussed in later sections in connection with the various specific problems involved. It appears that the general development proceeded in steps some 5 years apart. In 1955 to 1956 the significance of accommodating the image gas atoms to the tip temperature had been realized and led immediately to the operation of the microscope at cryogenic temperatures.[8] Fortunately at the same time the further pursuit of the field desorption process led to the discovery of field evaporation[9] which is next to field ionization the most significant physical effect on which field-ion microscopy is based.†

The concept of hopping image gas molecules, introduced in 1956,[10] turned out to be fruitful for all further theoretical considerations as well as the immediate improvements of the imaging procedures. Only at this stage was true atomic resolution of large sections of the specimen achieved. Direct visualization of the atomic crystal lattice and its defects was a reality. Vacancies, interstitials, and impurity atoms were seen as individual entities, and dislocation cores, slip bands, and cold-worked structures were revealed in intimate detail. By 1960 most of the present day experimental techniques were developed or initiated to give high-quality images of the lattice defects. Only then, perhaps helped by the appearance of a first detailed description of the principles and techniques involved,[11] did field-ion microscopy begin to be taken up seriously at various other places outside the laboratory of the author. During the last 5 years progress was made in the refinement of the theory,[12–14] in the more general recognition of the possibilities and limitations of the technique,[15] and in the increasing application of this research tool in electron physics of surfaces, in physical metallurgy, and potentially in molecular biophysics.

1.2   Basic Principles of the Field-Ion Microscope

The microscope was developed from its forerunner, the field-electron microscope,[4] which in its simplest form consists of the metallic specimen shaped to a finely pointed needle tip as a cathode and opposite to it a fluorescent screen as an anode, both mounted in a highly evacuated glass tube. With sufficient applied voltage, the field at the emitter reaches 30 to

---

† The term *field evaporation* and numerical data for room-temperature evaporation are mentioned in M. Drechsler and G. Pankow, *Proc. Intern. Conf. on Electron Microscopy, London,* 1954. However, these items were not presented at this conference, but were added to the proof after Dr. Drechsler had received a preprint of the *Phys. Rev.* article[9] in Sept., 1955. The *Proceedings* appeared late in 1956.

50 MV/cm, and the electrons tunneling out in normal direction to each surface element of the hemispherical tip cap radially project the specimen surface onto the fluorescent screen. The magnification is approximately equal to the ratio of screen distance to tip radius and can easily be made a million diameters, while the resolution limit is about 25 Å.

In the field-ion microscope the essential features are the same, but this time the specimen tip is usually of a smaller radius and kept at a higher, positive potential to produce a field of the order of magnitude of 500 MV/cm. The image information is carried from the tip surface to the screen by radially projected positive ions. The magnification is up to several million diameters, and the resolution often between 2 and 3 Å. The ions are not emitted from the specimen but are produced in its immediate proximity[16,17] by field ionization of the imaging gas which is introduced into the microscope tube at a pressure of a few millitorr, low enough to provide sufficient free path to let the ions travel to the screen without disturbing collisions. Two quite basic technical details are the provision for cooling the tip by heat conduction through its leads from a "cold finger" filled with a cryogenic liquid,[8] and a flat screen, which is required for photographing the weak images with a high aperture objective having a small depth of focus. A kind of heat shield around the tip, usually at screen potential, not only cools the gas arriving at the tip but also serves simultaneously as a means of restricting the volume containing the electric field, which is a useful design detail in a gas-filled high-voltage tube.[18] Another important technical detail in the early development was the use of a demountable microscope which allowed easy and fast tip replacements. Without this scheme, which with He or Ne as imaging gases still ensures atomically clear surfaces in spite of modest vacuum conditions,[18] the exploration of the many possibilities of field-ion microscopy[11] would have been much slower.

Under proper operating conditions the applied voltage is chosen so that ionization occurs only in the exceptionally strong field regions above the protruding atoms of the specimen surface or above the approximately circular edges of closely packed net planes. Accelerated by the extremely high field normal to the hemispherical surface, each of its protruding atoms sends a narrow beam of ions to the screen. The angular width of the beam, mostly determined by the random lateral velocity component of the ions and changing inversely with the tip radius, is as narrow as $\frac{1}{10}$ of 1°. Thus the total ion image, typically encompassing about two-thirds of a hemisphere, can be quite sharp and finely detailed if a large tip radius, in practice up to 2000 Å, is used. On the other hand, since at low tip temperature the resolution does not improve much with a smaller tip radius, the images of small tips, having radii down to below 100 Å, appear quite blurred owing to their unnecessary overmagnification. Fairly independent of the tip radius is the strength of the ion beam coming from a single surface atom, i.e., under practical conditions some 10^5 ions/sec or an ion current in the 10^{-14} A range.

As the specimen tip is simultaneously the image quality-determining "lens," some introductory remarks about shaping the tip are in order. Specimens are normally prepared from small cylindrical or rod-shaped samples, mostly in the form of fine wires, vapor-grown whiskers, or machined rods, by using chemical or electrochemical̄ etching and polishing to form a conical needle shape ending in the extremely sharp point with dimensions well below the range of an optical microscope. In field-electron microscopy the technique of finishing the emitter to a high degree of perfection by annealing the tip was found to be most useful. At temperatures above one-half or two-thirds of the melting point, surface migration of most metals becomes fast enough to rearrange the surface, which then approaches a shape of minimum free surface energy. The resulting frozen-in annealed end form, consisting of flat low-index crystal planes connected by smoothly curved intermediate regions, is regular enough for the limited resolution of the field-electron microscope but not for the field-ion microscope. Certainly, field-ion microscopy would not have reached its present capabilities if the effect of field evaporation[9] had not been discovered as a means to prepare essentially perfect tip surfaces. In the field-ion microscope the field at a crudely prepared tip is made so high that metal atoms evaporate from the surface even at cryogenic temperatures. As the field is gradually increased, evaporation of the most protuberant asperities occurs first because of the exceedingly large local field. This process continues as long as the voltage is kept high enough, and results in a field evaporation end form which is atomically smooth and crystallographically as perfect as the bulk material of the specimen. Once the end form is established, field evaporation can be continued at a well-controlled rate, which gives a welcome opportunity to "dissect" the specimen to explore the internal structure by "bringing it to the surface" for inspection.

The original image force theory of field evaporation,[9] only just recently refined,[14] served as a useful guide for the selection of materials suitable for field-ion microscopy. By 1958 images of some 18 metals and of carbon had been observed,[18] and the possibility of increasing the list of materials by going to other imaging gases such as neon, hydrogen, and deuterium had been explored.[19] The development of the standard microscopical techniques was essentially accomplished with the introduction of *in situ* treatments of specimens, such as cathode sputtering, α-radiation, cyclic field stressing for fatigue studies, and field stressing at elevated temperatures to investigate· yield phenomena.[11,18] The color printing technique[10] helped to pinpoint *in situ* surface changes. Taken together with the direct counting of vacancy concentration[20] and the imaging of defect structures containing slip bands and twin boundaries,[21] the possibilities of field-ion microscopy as a tool for metallurgical research was clearly established by about 1960. Since then, detailed work has been taken up at a number of laboratories all over the world. Some of the recent advances in the author's laboratory are the concept of field-stabilized surface sites and their interpretation as artifacts,[15] the successful application of the image intensifier for cinematographic recording of transient helium and neon ion images of unstable surfaces,[22] and the practical use of hydrogen promotion for imaging the nonrefractory transition metals which resulted from a closer investigation of gas–surface interactions.[23]

References

1. G. Gamow, Z. Physik 51: 204 (1928).
2. R. H. Fowler and L. Nordheim, Proc. Roy Soc. (London) A119: 173 (1928).
3. J. R. Oppenheimer, Phys. Rev. 31: 67 (1928).
4. E. W. Müller, Z. Physik 106: 541 (1937).
5. E. W. Müller, Naturwissenschaften 29: 533 (1941).
6. E. W. Müller, Z. Physik 120: 270 (1943).
7. E. W. Müller, Z. Physik 131: 136 (1951).
8. E. W. Müller, Z. Naturforsch. 11a: 87 (1956); also J. Appl. Phys. 27: 474 (1956).
9. E. W. Müller, Phys Rev. 102: 618 (1956).
10. E. W. Müller, J. Appl. Phys. 28: 1 (1957).
11. E. W. Müller, Advances in Electronics and Electron Physics, Vol. XIII, Academic Press (New York), 1960, pp. 83–179.
12. R. Gomer, Field Emission and Field Ionization, Harvard University Press (Cambridge, Mass.), 1961.
13. M. J. Southon, Thesis, Cambridge, England, 1963.
14. D. G. Brandon, Surface Sci. 3: 1 (1965).
15. E. W. Müller, Surface Sci. 2: 484 (1964).
16. M. G. Inghram and R. Gomer, J. Chem. Phys. 22: 1279 (1954).
17. E. W. Müller and K. Bahadur, Phys. Rev. 102: 624 (1956).
18. E. W. Müller, Proc. 4th Intern. Conf. Electron Microscopy, Berlin, 1958, Vol. 1, Springer (Berlin), 1960, p. 820.
19. E. W. Müller, Ann. d. Physik 20 [6]: 316 (1957).
20. E. W. Müller, Z. Physik 156: 399 (1959).
21. E. W. Müller, Acta Met. 6: 620 (1958).
22. S. B. McLane, E. W. Müller, and O. Nishikawa, Rev. Sci. Instr. 35: 1297 (1964).
23. E. W. Müller, S. Nakamura, O. Nishikawa, and S. B. McLane, J. Appl. Phys. 36: 2496 (1965).

Chapter 2

FIELD EMISSION AND FIELD IONIZATION

M. J. Southon†

2.1 Introduction

Both the field-ion microscope and the field-emission microscope are projection microscopes, in which the image is formed by radiation originating from a sharply pointed specimen and diverging from it to strike a fluorescent screen in a vacuum chamber. In the field-emission microscope the image is formed by electrons escaping from the negatively charged specimen by quantum-mechanical tunneling under the influence of an intense electric field. In the field-ion microscope the specimen is positively charged and the vacuum vessel contains a gas at a low pressure; the image is formed by the positive ions produced near the specimen's surface when electrons tunnel from their parent gas atoms into the specimen.

Since the field-emission microscope was both chronologically and conceptually the forerunner of the field-ion microscope, this paper will first describe briefly the field emission of electrons and its application in the field-emission microscope. The phenomenon of field ionization and the process of image formation in the field-ion microscope will then be described in more detail.

2.2 Field Emission

Thermionic emission, the photoelectric effect or photo emission, and secondary emission are processes in which the energy necessary to release an electron from a solid, the work function ϕ, is supplied respectively by phonons, by photons, or by more energetic particles. In field emission, however, the surface potential barrier is reduced by an applied electric field to such an extent that electrons may escape from a solid by tunneling through the barrier without change of energy (Fig. 2.1).

Although field emission from semiconductor surfaces has received extensive attention,[1,2] only the more straightforward case of emission from

† Assistant Director of Research, Department of Metallurgy, University of Cambridge, Cambridge, England.

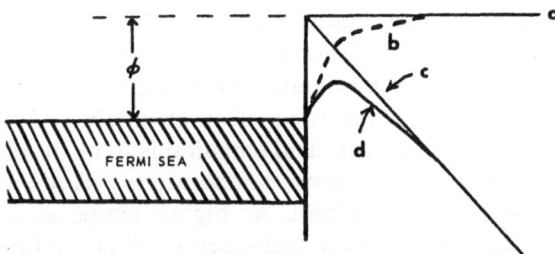

Fig. 2.1. Potential diagrams for an electron at a metal surface (*a*) in zero field; (*b*) rounding of the potential barrier by the image potential; (*c*) neglecting the image potential, an applied field produces a triangular barrier; (*d*) reduction of the triangular barrier by the image potential.

metals will be considered here. A qualitative criterion for the occurrence of field emission can be derived[2] from the uncertainty principle in the form $\Delta x \cdot \Delta p \sim \hbar$. Tunneling will be expected to occur if the momentum uncertainty Δp corresponds to the barrier height ϕ, $\Delta p \sim (2m\phi)^{\frac{1}{2}}$, and the uncertainty in position Δx corresponds to the barrier width ϕ/Fe. The Heisenberg relation then shows that the applied field strength F necessary to produce field emission must be of the order of $(2m\phi^3)^{\frac{1}{2}}/\hbar e$. For a clean tungsten surface for which $\phi \sim 4.5$ eV, the required value of F is 5×10^7 V/cm.

More accurately, the probability D that an electron incident on the surface barrier will tunnel through it can be calculated by the Wentzel–Kramers–Brillouin method. For the triangular barrier of Fig. (2.1) of height ϕ and width ϕ/Fe, D is found to be proportional to $\exp\left[-(2m/\hbar^2)^{\frac{1}{2}}\phi^{\frac{3}{2}}/Fe\right]$ for an electron at the Fermi level. A more realistic model of the surface potential barrier includes the image potential $-e^2/4x$ for an electron distant x from a conducting surface, due to the attraction of the electron by its positive electrostatic image in the surface. The effect of the image potential, known as the *Schottky effect*, is to reduce the height of the triangular barrier by an amount $(e^3F)^{\frac{1}{2}}$. An expression for the field-emitted current density J may then be obtained by integrating the barrier-penetration probability D over the distribution of energy of the incident electrons. Such an expression was first obtained by Fowler and Nordheim.[3] The well-known Fowler–Nordheim equation with the first-order image correction is

$$J = 6.2 \times 10^{-6} \frac{(\mu/\phi)^{\frac{1}{2}}}{\alpha^2(\phi + \mu)} F^2 \exp\left(-6.8 \times 10^7 \phi^{\frac{3}{2}} \frac{\alpha}{F}\right) \qquad (2.1)$$

where μ is the Fermi energy measured relative to the bottom of the conduction band, μ and ϕ are in units of electron volts, F is in volts per centimeter, and J is in amperes per square centimeter.[2] The image correction term α, discussed by Gomer,[2] is a function of F and ϕ and is some 10 to 20% less than unity under practical conditions. If $\phi \sim 4.5\,\text{eV}$, $F \sim 5 \times 10^7\,\text{V/cm}$, the predicted current density is approximately $1000\,\text{A/cm}^2$.

Although the Fowler–Nordheim equation given above applies only at $0°\text{K}$, the temperature dependence of field emission is weak at moderate temperatures; the current density at room temperature exceeds the low temperature value by only a few percent. At higher temperatures, however, above about $1000°\text{K}$, electrons in the high-energy tail of the Fermi distribution make a substantial contribution to the emission current. The region of T–F emission, in which both field emission and thermionic emission are important, has been extensively investigated in connection with the development of stable field-emission electron sources for use at very high current densities.[4]

The experimental confirmation of the validity of this theory of field emission, discussed, e.g., by Good and Müller,[5] is based largely upon the verification of the Fowler–Nordheim equation but also on the agreement between the calculated and observed energy distributions of the field-emitted electrons. The total-energy distribution has been calculated by Young[6] and measured by Young and Müller[7] and others (see Ref. 8).

2.3 Field-Emission Microscopy

Although the emission of electrons from surfaces subject to intense electric fields had been reported as early as 1897,[9] it was not until 1938 that the last major discrepancies between theory and the observed behavior of field emitters were removed. Measurement of field-emission currents had always been complicated by the fact that emission occurred predominantly in the enhanced fields acting at small irregularities or asperities on the emitter surfaces. In 1937 Müller showed that point emitters smoothed by surface migration of metal atoms at elevated temperatures in a good vacuum gave steady and reproducible emission currents in good agreement with theory. Furthermore, in the same year Müller took the most important step of placing a fluorescent screen in the path of the emitted electron beam so that any variations of current density across the beam could be observed visually. This was the first field-emission microscope.[10,11]

A field-emission microscope tube is shown schematically in Fig. 2.2. The point emitter is mounted on a filament, which may be heated resistively, facing a fluorescent screen coated onto the conducting inner surface of the glass vacuum tube. The field F at the surface of a point emitter is related to the voltage V applied to it with respect to the earthed fluorescent screen by the approximate expression $F = V/kR$, where R is the mean radius of curvature of the emitter surface and the constant $k \sim 5$. A voltage of 1.5 to 3.0 kV must therefore be applied to an emitter of 1000 Å radius to generate

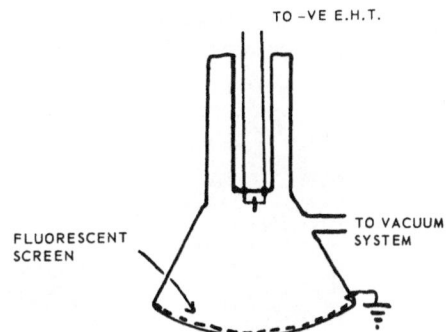

Fig. 2.2. Schematic field-emission microscope tube.

the fields of 3 to 6 × 10⁷ V/cm commonly used in field-emission microscopy;
the radii of curvature of emitters are usually in the range of 1000 to 10,000 Å.

The emitted electrons leave the surface of the emitting point in essentially radial directions to form a greatly enlarged image on the fluorescent
screen. The magnification is of the form $R'/\beta R$, where R' is the emitter-screen
distance and R the emitter radius. The constant β depends on the geometry
of the microscope tube and typically has a value near 1.5. The magnification
for an emitter of 3000 Å radius 10 cm from the fluorescent screen is thus
$\sim 2 \times 10^5$.

The origin of contrast in the field-emission image, a typical example of
which is shown in Fig. 2.3, may be seen from the Fowler–Nordheim relation,
equation (2.1). The emitted current density is seen to be an exponential
function of $\phi^{\frac{3}{2}}/F$, and the field-emission image therefore sensitively displays
any variations in the local values of work function or field strength across
the emitter surface. The variation of field strength over the surface reflects
the variation of the effective local radius of curvature and consists of a
relatively large scale variation due to the over-all shape of the emitter, which
is generally of only secondary interest in field-emission microscopy, and
shorter range variations due to the detailed topography of the surface. The
dominant variable determining the contrast in the field-emission image,
however, is the work function, which varies with crystallographic orientation over the curved surface of the emitter, usually a single crystal because
of its small size, and is also sensitive to changes in composition due to
impurities adsorbed from the gaseous phase or diffused to the surface from
the emitter interior.

The work function ϕ consists of two components. The first, which may
be termed the *absolute work function*, is simply the difference between the
total energy of an electron at the Fermi surface in the metal and the energy
of a stationary electron at infinity outside the metal, and it is clearly a property
of the bulk metal. The work function measured experimentally, however, is

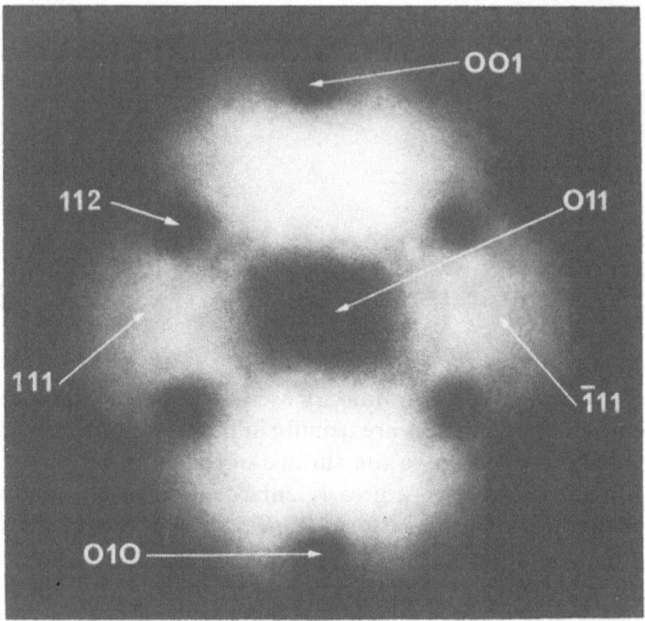

Fig. 2.3. Field-emission micrograph of a clean annealed tungsten emitter. (Courtesy of Dr. A. J. Melmed, U.S. National Bureau of Standards.)

usually the energy required to move an electron over the surface potential barrier to some point just outside the surface and includes a contribution due to the change in electrostatic potential through the electrical double layer present at a surface. The strength of the double layer depends on the nature of the individual surface, and the measured work function is therefore anisotropic. It should be noticed that the anisotropy of the work function implies the presence of a local field acting around the surface of a metal crystal; the precise value of the measured work function depends on the relative magnitudes of the local fields and the extracting field used in a particular experimental arrangement; this topic has been fully discussed by Herring and Nichols.[12]

The strength of the double layer at a clean metal surface depends on the way in which the electron charge density falls to zero outside the surface (see, e.g., Refs. 13 and 14). Smoluchowski[15] has given a clear account of the physical situation and has shown that the slow decay of electron density through a relatively close-packed surface leads to a double layer with negative charge outward, which tends to increase the work function, whereas the spreading of the electron cloud between the protruding atoms of a higher-index plane gives a double layer with positive charge outward, in agreement with the observation that the work functions of close-packed planes are larger than those of more open surfaces. Adsorption of gas on a clean surface causes a redistribution of charge in such a way that an additional dipole moment may be associated with each adsorbed particle[2]; the dipole moment, and hence the change in work function due to adsorption, may be of either sign.

The resolution with which the field-emission image displays local variations in work function or field at the emitter surface is limited by two effects; firstly, the spread of transverse momenta of the electrons leaving the emitter surface and, secondly, diffraction of electrons passing through the emitter surface or, an alternative view of the same effect, the limit imposed by the uncertainty principle on the width of the wave packet representing an electron of finite transverse momentum. The quantitative calculation of the resolution will be indicated later for the field-ion microscope. For the field-emission microscope the resolution δ in distance on the emitter surface is found to be

$$\delta = 2.62 \times 10^{-4} \beta R^{\frac{1}{2}} \left(\frac{1.16}{\beta V^{\frac{1}{2}}} + \frac{1}{k\alpha\phi^{\frac{1}{2}}} \right)^{\frac{1}{2}} \tag{2.2}$$

where δ and R are in centimeters and V in volts.[2,5] The first term represents the contribution of the transverse momenta, and the second, the diffraction effect. For an emitter of 10^{-5} cm radius, $\phi = 4.5$ eV at a field of 5×10^7 V/cm, the predicted resolution is of the order of 20 Å, in agreement with observation. Although for a very small emitter or a sharp protuberance on an emitting surface the resolution could be less than 10 Å. The field-emission microscope is therefore not in general capable of resolving atomic detail.

Before considering the applications of field-emission microscopy, it is useful to note some practical aspects of the technique. The current densities obtained from emitters of 10^3 to 10^4 Å radius at applied voltages of 1 to 20 kV typically yield total currents of between 10^{-8} and 10^{-5} A; the corresponding images on the fluorescent screen are then bright enough for visual observation without dark adaptation. Currents of this order clearly prohibit the use of insulators and poor semiconductors as field emitters, since the applied voltage would be dropped along the shank of the emitter rather than across the vacuum. A further restriction on emitter materials is imposed by the mechanical stress $F^2/8\pi$ at the emitter surface due to the applied field F. A field of 4.5×10^7 V/cm gives rise to a stress of about 10^9 dynes/cm^2, sufficient to fracture emitters of weak or brittle materials.

Field-emission microscopy, like other techniques for the study of surfaces, necessitates the use of ultra-high vacuum. Emitter surfaces are normally cleaned by heating to a temperature sufficient to desorb adsorbed impurities without permitting the emitter to be blunted by surface diffusion.[2,4,5] Considerable difficulties can arise when neither thermal desorption nor field desorption will yield a clean emitter surface. Ambient pressures in the 10^{-10}-torr range are required to maintain a clean surface for the duration of an experiment and also to minimize damage to the emitter by sputtering due to residual gas ions.

Thus the technique of field-emission microscopy permits the observation of small variations in the work function and local curvature of a point emitter with a resolution of about 20 Å under ultrahigh vacuum conditions. The applications of the technique have been described in the literature (e.g., see Refs. 2, 5, 8, 16) and need only be briefly summarized here. The most fruitful application of field-emission microscopy has been to the study of adsorption phenomena,[2,17] in which the small changes of work function due to adsorption may be observed as a function of crystallographic orientation, temperature, adsorbate coverage, and field strength. The measurement of work function usually employs the Fowler–Nordheim plot, in which the variation of emission current I with applied voltage V is plotted in the form of $\log(I/V^2)$ *versus* $1/V$ to yield a straight line whose slope is proportional to $\phi^{\frac{3}{2}}/F$ [equation (2.1)]. Absolute values of ϕ and F may be determined if a second function of (ϕ, F) is measured experimentally; Young and Müller[18] have suggested measurement of the energy distribution of the emitted electrons or the temperature dependence of the emission current. Recent field-emission studies of adsorption have emphasized the need to measure work functions of individual crystal planes.[8,19]

Such phenomena as oxidation and corrosion, diffusion of impurities to surfaces, and the formation of surface phases may also be studied through the medium of work-function variations, whereas surface diffusion of the emitter material and estimation of surface energies, studies of whisker growth and surface geometry rely on the influence of the local field at the emitter surface on the image contrast.

2.4   Field-Ion Microscopy

From the calculation of the resolution of the field-emission microscope it is evident that, if an emission image could be formed by particles more massive than electrons, a substantially improved resolution could be expected. A standard technique in field-emission microscopy is the desorption of adsorbed material either by thermal activation or by the process of field desorption when a positive potential is applied to the emitter. The field-desorbed atoms or molecules are driven from the emitter surface as positive ions, but early attempts to form images with the desorbed ions were unsuccessful, partly because of the difficulty of replenishing the adsorbed layer at a rate sufficient to permit the formation of a visible image on the fluorescent screen.[20]

However, in 1951 Müller[21] experimented with the operation of a conventional field-emission microscope tube with the emitter at a positive potential in the presence of a low pressure of hydrogen instead of the usual high vacuum. Under these conditions a visible hydrogen-ion image was in fact formed, and the resolution of the image, although not sufficient to show detailed atomic arrangements, had improved to between 5 and 10 Å. In 1954 mass spectrometric analysis of the image-forming ions by Inghram and Gomer[22] showed that gas molecules could be ionized in the very strong fields near the emitter surface without the necessity of prior adsorption, a process known as *field ionization*. In 1956 a further crucial experiment, again in Müller's laboratory, showed that ionization of helium gas over an emitter cooled to a low temperature yielded images in which the arrangement of the individual atoms of the emitter surface could plainly be seen, with a resolution of 2 to 3 Å.[23] The fields necessary for the ionization of hydrogen and helium range from 2 to 6×10^8 V/cm, or 2 to 6 V/Å, an order of magnitude larger than the negative fields required for electron emission. Field-ion emitters therefore have radii from 100 to 1000 Å to give images in the range of 5 to 30 kV applied voltage. Gas pressures are normally in the 10^{-3}-torr range, to give maximum image-brightness without loss of resolution by ion–gas atom scattering.

The processes involved in the formation of field-ion images will now be discussed in more detail.

2.5 Field Ionization

Inghram and Gomer[22] measured the mass and energy distribution of the ions produced from a number of gases in a field-ion microscope. The similarity of the observed ionization characteristics of hydrogen and deuterium suggested that tunneling of ions was not involved since such a process would be strongly dependent on the mass of the tunneling particle. The ions observed also included species which were unlikely to be adsorbed on the emitter, which suggested that adsorption was not a necessary part of the ionization process. The observation that the energy distributions of the emitted ions were sharp to within the limited resolution of the apparatus at low fields but spread toward the low-energy side at high fields suggested that ionization occurred in space near the emitter surface at low fields and at increasing distances from it in higher fields.

Figure 2.4a shows the potential energy of an electron near the parent ion in field-free space. Figure 2.4b shows the effect of an applied field; ionization can now occur from the ground state of the atom if the electron tunnels through the potential barrier along the path AB. Figure 2.4c shows

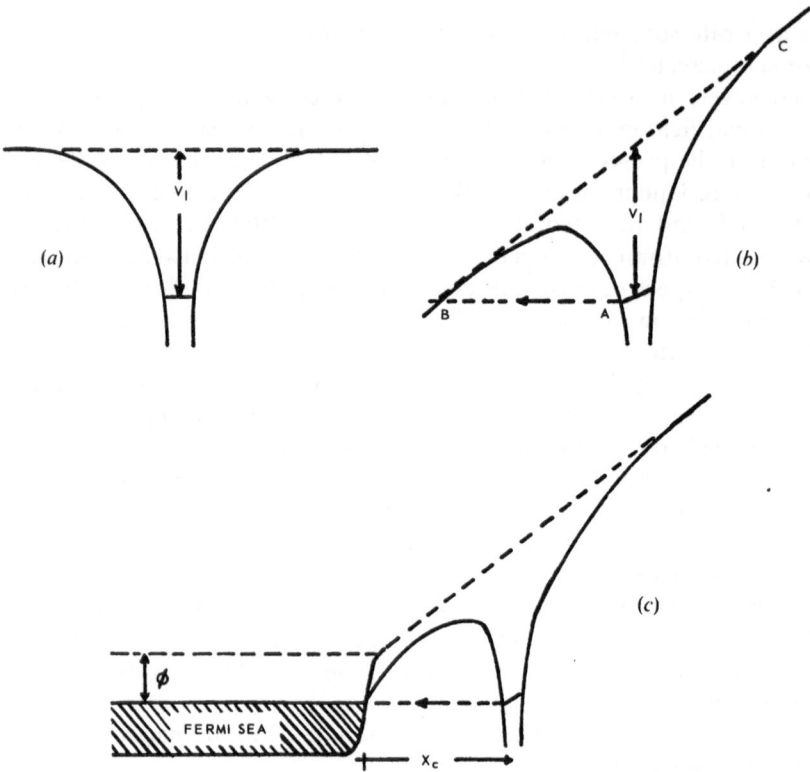

Fig. 2.4. Potential diagrams for an electron in field ionization; (a) the potential due to the parent ion in zero field and (b) in a field BC; (c) the potential in an applied field near the surface of a metal of work function ϕ.

that, when the atom is near the surface of a conductor, the potential barrier is further reduced by image forces which attract the electron to the image of the ion–electron dipole in the conducting surface. In Fig. 2.4c the gas atom is shown at a critical distance x_c from the emitter surface; if the atom moves closer to the surface, the electron energy level in the atom falls below the Fermi level in the metal and the tunneling probability is greatly reduced because of the low density of vacant states to which the electron may tunnel.

The barrier-penetration probability D for the tunneling electron may be calculated from the WKB approximation and is of the form

$$D = \exp\left[-\left(\frac{8m_e}{\hbar^2}\right)^{\frac{1}{2}} \int_{x_2}^{x_1} (V - E)^{\frac{1}{2}}\, dx \right] \tag{2.3}$$

where m_e is the electron mass and E is its total energy. The potential energy

V of the electron at a distance x from the emitter surface is not known accurately for small x but is approximately given by

$$V(x) = \frac{-e^2}{|x_n - x|} + Fex - \frac{e^2}{4x} + \frac{e^2}{x_n + x} \qquad (2.4)$$

in which the first term represents a coulomb attraction to an ion at a distance x_n from the emitter surface and the last two terms are image potentials due to the electron and ion images, respectively.

The lifetime τ_i with respect to ionization of an atom distant x_n from the emitter surface is given by

$$\tau_i(x_n) = (vD)^{-1} \qquad (2.5)$$

The frequency v with which the electron approaches the potential barrier may be simply estimated from the Bohr model of an atom or from the uncertainty principle. The probability dP_i that an atom will be ionized while traversing a distance dx normal to the emitter surface is

$$dP_i = \frac{dx}{v_r} \tau_i(x) \qquad (2.6)$$

where v_r is the velocity of a gas atom in a direction normal to the emitter surface. Müller and Bahadur[24] have calculated values of the differential ionization probability dP_i/dx, using a numerical integration of equation (2.3) for the potential of equation (2.4) and applying a correction for the three-dimensional nature of the tunneling process. Using also the simplifying assumption that v_r has the constant value $F(\alpha/m)^{\frac{1}{2}}$, where m and α are the mass and polarizability of the gas atom and F is the field strength at the emitter surface, an assumption whose significance will appear later, Müller and Bahadur obtained the results shown in Fig. 2.5. It can be seen that the ionization probability depends very strongly on the field strength and also increases sharply toward the emitter surface as the short-range image forces become important.

The curves of Fig. 2.5 terminate at the critical distance x_c shown in Fig. 2.4 and defined[20] by the equation

$$Fex_c = V_I - \phi - \frac{e^2}{4x_c} + \tfrac{1}{2}F^2(\alpha_A - \alpha_I) \qquad (2.7)$$

in which V_I is the ionization potential of the gas atom, $-e^2/4x_c$ is the ion–ion image potential, and the final term represents the difference in polarization potential energies of the gas atom and the resultant ion. Equation (2.7) may be simplified to give $x_c \cong (V_I - \phi)/Fe$, from which x_c is found to lie in the

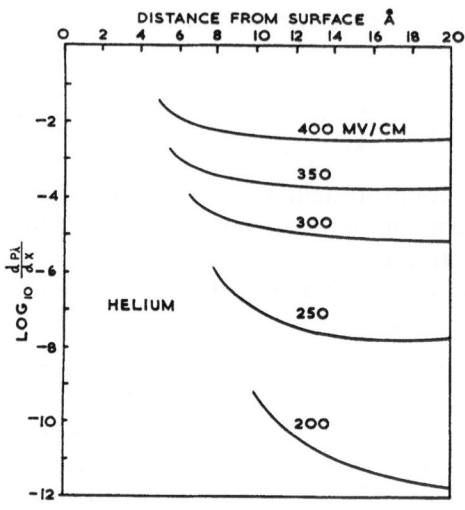

Fig. 2.5. Calculated values of the differential ionization probability dP_i/dx for a helium atom at various field strengths. (From Müller and Bahadur.[24])

range of 4 to 8 Å for most of the gases likely to be used in field-ion microscopy. This simple model of the field-ionization process is clearly unreliable at such short distances. In particular the assumptions of a smooth emitter surface, a one-dimensional tunneling process, and the classical form of the

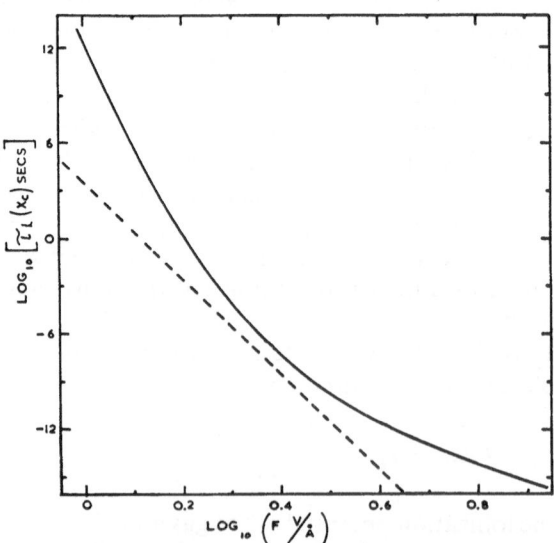

Fig. 2.6. Log–log plot of $\tau_i(x_c)$, the characteristic time for ionization at the critical distance, against the applied field F. The dashed line shown for comparison has a slope of -30.

image potential are clearly invalid, but in the absence of a more realistic model of the relevant gas–surface interaction further refinement of the simple model seems unjustified.

The parameter $\tau_i(x_c)$, the ionization lifetime of a gas atom at the critical distance from the emitter surface, is of importance as a measure of the minimum of τ_i. Figure 2.6 shows the variation of $\tau_i(x_c)$ with the field strength F at the emitter surface obtained by the present author[25] by graphical integration of equations (2.3) and (2.4) for helium ionization over a tungsten emitter. The result emphasizes the very strong dependence of ionization lifetime on field strength. As in the calculations of Müller and Bahadur, the variation of field strength with distance from the emitter surface has been neglected, but it has been shown[25] that the variation of τ_i with x is due predominantly to image forces except for emitters of a radius less than about 150 Å.

Simple theory thus predicts that the probability of ionization for a gas atom moving near the surface of an emitter subject to a strong positive electric field will depend sensitively on the magnitude of the field and will increase sharply toward the emitter surface, so that ionization will occur predominantly near the critical distance x_c, a few Å from the surface.

2.6 Field-Ion Energy Distribution and Current–Voltage Characteristics

The experimental evidence relating to the simple theory of field ionization may be summarized under three headings: the observed characteristics of field-ion images, measurements of the energy distributions of the field ions and measurements of the ion current-voltage characteristics.

The characteristics of field-ion images are in fact of little value in the quantitative verification of the theory of field ionization. Little is known about the variation of electron density near a field-free surface of a real material, and little can be said on theoretical grounds about the likely modulation of the field strength at the surface of a charged emitter by the atomic details of the surface. It is therefore not possible to deduce theoretically the variation of current density to be expected across a field-ion image or to account for differences in the threshold fields at which ionization will start to occur over different crystal planes. While the general characteristics of field-ion images provide broad confirmation of some aspects of the theory outlined above, it is in general more useful to make deductions from the theory of field ionization about the detailed nature of the emitter surface rather than *vice versa*.

The original measurements of field-ion energy distributions by Inghram and Gomer[22] provided an early basis for the understanding of the formation of field-ion images, and the more recent and refined work of Tsong and Müller[26] gives a quantitative verification of the theory of field ionization for the case of low-temperature field-ion microscopy. Tsong and Müller used a retarding-potential technique based on the earlier electron-emission work of Young and Müller[7] to measure the energy distributions of field ions from a small region of a tungsten emitter with a resolution of 0.03 V, corresponding to a spatial resolution of 0.01 Å in a field of 3 V/Å. Agreement between the observed and calculated high-energy cutoff of the distributions, corresponding to ionization at x_c, was to within 0.2 or 0.3 eV for He, Ne, A, and H ionization, in striking confirmation of the theory. The minimum half-widths of the distributions were found to be between 0.5 and 0.7 eV for those gases, corresponding to ionization in a zone above the emitter surface of only a few tenths of an Å in depth. The half-widths increased slowly with field strength (Fig. 2.7), as the region in which the field strength exceeded the threshold value spread further from the emitter surface. Figure 2.8 shows a small increase in the distribution width with increasing temperature, which reflects the fact that not only the ionization lifetime but the distribution of normal velocity components v_r [equation (2.6)] of gas atoms before ionization affects the energy distribution of the field ions and the nature of the field-ion image, as discussed in the next section. Half-widths calculated from a one-dimensional approximation by Tsong and Müller[27] and by a more sophisticated method based on collision theory by Boudreaux and Cutler[28] are in good agreement with the observed values.

Measurements of the current–voltage characteristics of field-ion emitters also provide data which may be compared with the theory of field ioniza-

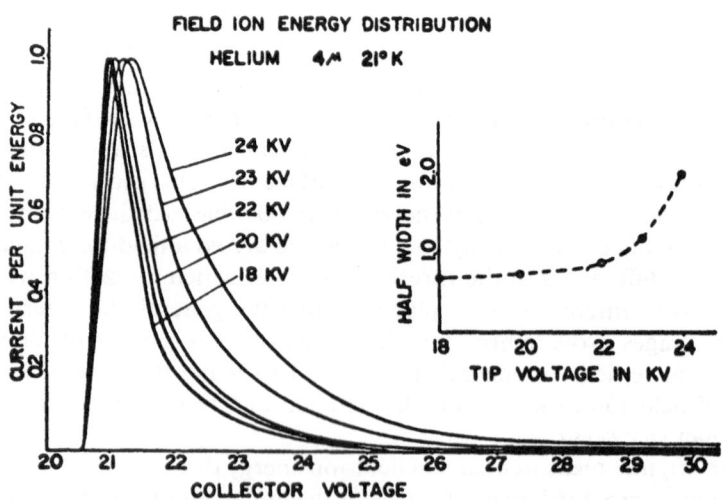

Fig. 2.7. Normalized energy distribution for helium ions obtained by Tsong and Müller.[26] The best image voltage 20 kV corresponds to a field of 4.4 V/Å.

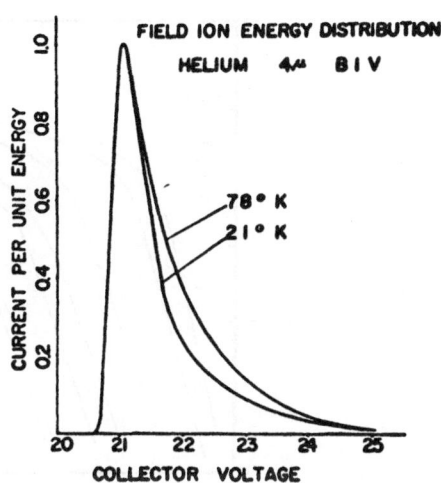

Fig. 2.8. The effect of temperature on the energy distribution of helium ions at the best image field. (From Tsong and Müller.[26])

tion. The variation of helium-ion current with applied voltage observed for a tungsten emitter by Southon and Brandon[29] is shown in Fig. 2.9. The upper limit of the curve at the point D marks the onset of field evaporation, the field-induced evaporation of ions from the parent surface to be discussed in Chapter 3. The ion current is seen to increase initially very sharply with applied voltage and field, typically as F^n with the index n between 20 and 40, as might be expected from the strong dependence of $\tau_i(x_c)$ on field strength shown in Fig. 2.6. The relatively slow increase of ion current with field in the region BCD in Fig. 2.9 suggests that above the point B the ionization probability has saturated at a value corresponding to the certain ionization

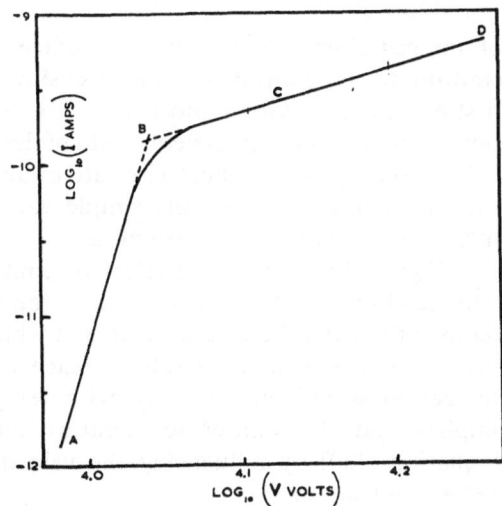

Fig. 2.9. The typical variation of helium-ion current with voltage for a tungsten emitter at 77°K (log–log plot). Emitter radius = 570 Å, helium pressure = 6×10^{-3} torr. The slopes of the linear portions AB and BD are 28.9 and 2.94 respectively.

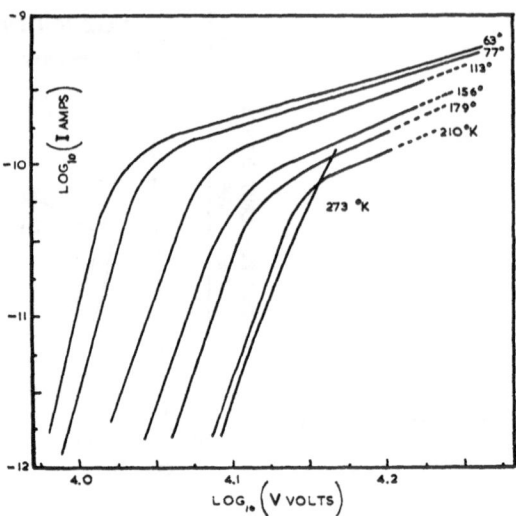

Fig. 2.10. The current-voltage characteristics of a tungsten emitter of about 630 Å radius at various temperatures and at a helium pressure of 6×10^{-3} torr. The upper limits of the curves for 63° and 77°K are at the evaporation voltages. At the remaining temperatures, current measurements were terminated well below the evaporation voltages in order to avoid changes of emitter radius due to field evaporation. The dashed extrapolations for temperatures from 113 to 210°K are terminated at voltages corresponding to calculated evaporation fields.

of any gas atom near the emitter surface but that the supply of gas to the emitting region continues to increase slowly with field strength. This effect, first explained by Good and Müller,[5] is due to the attraction of gas atoms, polarized in the inhomogeneous field of the emitter, to the region of strongest field at the emitter surface. The rate of arrival of gas atoms at the emitter surface therefore exceeds the simple gas kinetic value by a factor which increases slowly with field strength.

Figure 2.10 shows the effect of emitter temperature on the current-voltage characteristic.[29] The evaporation field falls slowly with increasing temperature, and the threshold field at which ionization starts to occur rises more strongly with temperature. Since the tunneling probability in field ionization is unlikely to vary with temperature, the latter observation implies that the emitter temperature affects the ionization probability [equation (2.6)] by influencing the velocity distribution of the gas atoms before ionization.

2.7 Gas Supply to a Field-Ion Emitter

It has been shown that the characteristics of field ionization depend not only on the tunneling probability but also on the detailed behavior, in particular the velocity distribution, of the gas atoms before ionization.

The potential energy U of a gas atom or molecule in an electric field F may be written as

$$U = -\tfrac{1}{2}\alpha F^2 - \mu F \tag{2.8}$$

where α is the polarizability and μ is any permanent dipole moment associated with the particle. It is useful to consider the state of the gas near the emitter when the applied field is too low to produce appreciable ionization. The concentration of gas atoms at the emitter is then $C_0 \exp(|U|/kT)$ if C_0 is the concentration in a field-free region, from Maxwell–Boltzmann statistics. The flux f of gas atoms near the emitter is therefore $f_0 \exp(|U|/kT)$, where the flux f_0 in zero field is $P/(2\pi mkT)^{\frac{1}{2}}$ at pressure P for gas atoms of mass m. The fraction of the atoms in this flux with a kinetic energy greater than U is $(1 + |U|/kT)\exp(-|U|/kT)$, so that the number of atoms leaving a unit area of the emitter per unit of time with sufficient kinetic energy to escape to infinity is simply $f_0(1 + |U|/kT)$, two exponential functions of $\pm|U|/kT$ having canceled. The flux of gas atoms that actually do escape from the emitter, however, is given by $f_0 \delta(1 + |U|/kT)$, where δ is the fraction of atoms with sufficient energy to escape that leave the emitter in a direction which leads to escape. (A charged particle leaving one plate of a parallel-plate condenser with a kinetic energy equal to the difference in potential energy between the plates will only reach the other plate if its initial velocity is normal to the plates.)

Since in the absence of ionization the flux of gas atoms escaping from the emitter is equal to the flux arriving, it follows that the incident flux to an emitter at a field F exceeds f_0 by an enhancement factor σ equal to $\delta(1 + |U|/kT)$. The fraction δ depends on the detailed nature of the force field surrounding the emitter and can only be evaluated for the potential of equation (2.8) for emitters of high symmetry. The author[25] has calculated the enhancement factors σ_s and σ_c for spherical and cylindrical emitters and has found

$$\sigma_s = \left(\frac{\pi \alpha F^2}{2kT}\right)^{\frac{1}{2}} + \mu F \tag{2.9}$$

$$\sigma_c = 2\left(\frac{\tfrac{1}{2}\alpha F^2 + \mu F}{\pi kT}\right)^{\frac{1}{2}} \tag{2.10}$$

for $\tfrac{1}{2}\alpha F^2 \gg kT$, σ_s and σ_c both tending to unity for small F.

Since the flux of gas atoms arriving at the emitter from field-free space is almost independent of the probability of ionization at the emitter and since the shape of a point emitter may be roughly represented as a hemispherical cap on a cylindrical shank, the gas supply f to a point emitter in a nonpolar gas will be approximately

$$f = f_0 \cdot \frac{3}{2}\left(\frac{\alpha F^2}{2kT}\right)^{\frac{1}{2}} \tag{2.11}$$

The quantity of $\alpha F^2/2kT$ typically lies between 10 and 100 in practical field-ion microscopy, so that the enhancement of the gas supply by polarization forces is substantial.

Equation (2.11) describes the flux of gas arriving from field-free space at a point on the emitter surface where the field is F. In order to estimate the supply of gas to the narrow zone above the apex of the emitter in which ionization occurs, it is necessary to consider the behavior of gas atoms in more detail. There is a small probability that a gas atom arriving at the emitter with kinetic energy of $\sim\frac{1}{2}\alpha F^2$ will be ionized on its first pass through the ionization zone, and a further small probability that it will be ionized on its second pass after reflection at the emitter surface. But it does not follow that any gas atom which survives these first two passes will escape from the emitter to infinity: whether an atom reflected from the emitter surface will escape or become trapped near the emitter depends both on its kinetic energy after reflection and on the direction in which it leaves the emitter surface.[25]

The kinetic energy with which a gas atom initially arrives at the emitter surface is the sum of the contribution due to attraction by the polarization forces and the original thermal energy of the atom, $\sim\frac{1}{2}\alpha F^2 + kT$. Clearly, if an atom loses an energy greater than its original thermal energy kT, its resultant *total* energy will be negative and it will be unable to escape from the emitter to infinity. Similarly, there is a finite probability that an atom which retains a positive total energy after reflection will be reflected in an unfavorable direction and will also be attracted back to the emitter and become trapped. On the basis of certain assumptions about the gas–surface interaction and, in particular, about the accommodation coefficient which describes the fractional energy loss of a gas atom in a collision with a surface, the author[25] has derived general expressions for the trapping probability of gas atoms initially incident at any point on the emitter surface.

In practical situations the accommodation coefficient is often of the same order of magnitude as $kT/\frac{1}{2}\alpha F^2$, the fraction of its energy that a gas atom must lose to become trapped. The trapping probability therefore depends fairly strongly on field strength and temperature over the practical range of these parameters, and the rate of arrival Z of gas atoms that become trapped and eventually reach the ionization zone must be obtained by integrating the product of the flux of gas atoms arriving from field-free space [equation (2.11)] and the trapping probability over the whole area of the emitter apex and shank. A calculation of this type by the author[25] gave results in reasonable agreement with measured ion current-voltage characteristics[29] and showed that a substantial fraction of the gas atoms contributing to a typical field-ion image initially strike the shank of the emitter, become trapped, and are then drawn to the ionization zone at the apex of the emitter by polarization forces.

The foregoing considerations permit the derivation of explicit expressions for the ion current leaving an emitter; this derivation will now be indicated briefly. Following Gomer's formulation of the problem,[2] the total ion current I may be expressed as

$$I = ek_i n_t \qquad (2.12)$$

where n_t is the number of gas atoms trapped at the emitter and k_i is a rate constant for ionization, related to $\tau_i(x_c)^{-1}$. In the steady state n_t is defined by

$$n_t = (k_i + k_e)^{-1} Z \qquad (2.13)$$

where Z is the rate of arrival of gas atoms at the emitter apex defined above and k_e is the rate constant for escape of gas atoms from the potential well created at the emitter by polarization forces. In general, therefore,

$$I = \frac{eZk_i}{k_i + k_e} \qquad (2.14)$$

At low field strengths where $k_i \ll k_e$, equation (2.14) becomes

$$I = \frac{eZk_i}{k_e} \qquad (2.15)$$

The rate constant k_e may be calculated since the process of thermally activated escape is to a good approximation simply the reverse of the supply and trapping process described above. Substitution leads to the expression

$$I = eAf_0\tau_i^{-1} \exp \frac{\frac{1}{2}\alpha F^2}{kT} \qquad (2.16)$$

for the ion current from an emitter of area A at low field strength.[2,25] The variation of ion current with field is determined jointly by the strong field dependences of τ_i and the Boltzmann factor, and the influence of temperature on the threshold field for ionization is also accounted for.[25] Equation (2.16) corresponds to relatively infrequent ionization of the gas atoms present in

a large concentration $\sim C_0 \exp \dfrac{\frac{1}{2}\alpha F^2}{kT}$ at the emitter surface.

At high fields where $k_i \gg k_e$, equation (2.14) reduces to

$$I = eZ \tag{2.17}$$

corresponding to the presence of a low concentration of gas at the emitter surface and the prompt ionization of all gas atoms arriving at a rate determined by the supply function Z.

At intermediate fields where k_i and k_e have comparable values, it is necessary to consider the detailed form of the rate constants, as has been done by Gomer[2] and the author.[25] The results of equations (2.16) and (2.17), however, are sufficient to account in general terms for the ion current measurements of Figs. 2.9 and 2.10, and similar considerations to the above must be used in calculating the energy distributions of the emitted ions.

An important feature of the behavior of gas atoms before ionization is that the accommodation coefficient describing a gas–surface collision is likely to be small. An incident gas atom that becomes trapped will therefore strike the emitter surface between 10 and 100 times before its kinetic energy falls to the value appropriate to the emitter temperature. For the incident atoms that ultimately escape from the emitter it can be shown[25] that escape will occur either after the first one or two collisions with the emitter surface or after a very much larger number for an atom that becomes accommodated to the emitter temperature. Since ionization of the energetic atoms during their first few passes through the ionization zone only occurs at very high field strengths, the gas atoms that are ionized under normal conditions will all be in thermal equilibrium at the emitter temperature.

Such gas atoms will continue to move over the emitter surface, leaving the surface with an energy of about kT after reflection and bouncing to a height above the emitter surface determined by the strength of the polarization forces before being attracted back to the emitter. This concept of "hopping" was first introduced by Müller.[30] It can be shown that the average hop height is approximately $RkT/4\alpha F^2$ above an emitter surface of radius R, about $\frac{1}{100}$ of the tip radius for helium gas at 77°K and $\frac{1}{500}$ at 21°K. Since this height may be comparable with or smaller than the critical distance x_c, an effective activation energy term must be introduced into the expression for k_i to describe the probability that a gas atom will hop to a sufficient height to enter the ionization zone.[2] This effect is partly responsible for the observed variation of the threshold field for ionization with temperature[25] and with emitter radius.[25,31]

2.8   Resolution of the Field-Ion Microscope

Following the conventional ideas of object and image in optics, the object to be resolved in the field-ion microscope may be taken for most purposes to be the atomic structure of the emitter surface and the image to be the pattern which would be recorded by a perfect detector in place of the fluorescent screen. The latter definition dismisses many of the practical problems of field-ion microscopy so that, with the further assumption that scattering of the field-ions by the parent gas is negligible, the remaining variables determining the resolution of the field-ion microscope may be considered under three headings: firstly, the extent to which the atomic structure of the emitter surface is reproduced as a spatial modulation of the electric field strength through a surface x_c above the emitter, where ionization predominantly occurs; secondly, the extent to which this modulation of the field strength produces a modulation of the ion current density leaving the emitter; and thirdly, the extent to which the spatial modulation of the ion current density above the emitter is retained in the ion beam reaching the fluorescent screen.

Little can be said from first principles about the first factor. The extent of the electric field modulation above a surface clearly depends on the spacing and arrangement of the atoms in the underlying crystal plane. It may prove to be possible to deduce information about the field modulation from the ion image characteristics or from observations of field evaporation. It is also likely that the modulation of field strength increases with the applied field. A field strength of 5 V/Å corresponds to a surface charge deficiency of one-third of an electron per atom, and the critical distance x_c is also a function of field [equation (2.7)].

The second factor, the extent to which the field modulation is converted to a modulation of the emitted ion current density, is likely to be very favorable to good resolution since the probability of ionization depends sensitively upon the local field strength (Fig. 2.6).

The third factor, the extent to which ion current density modulations are retained in the ion beam reaching the emitter, is amenable to calculation if it is assumed that the electric lines of force near the emitter surface are strictly radial. Gomer[2] (see also Refs. 5 and 20) has shown that the transverse displacement $D/2$ of an ion at the fluorescent screen due to an initial tangential velocity component v_t at the emitter is $2R'v_t/v_f$, where R' is the emitter-screen distance and v_f is the terminal velocity of the ion. Therefore, D corresponds to a distance δ at the emitter of $4R\beta(E_t/eV)^{\frac{1}{2}}$, where E_t is the energy corresponding to the velocity component v_t and V is the applied voltage.

Diffraction effects also contribute a loss of resolution. A particle localized within a distance δ_0 parallel to the emitter surface has, from the Heisenberg principle, a transverse velocity $v_t \sim \hbar/2m\delta_0$. An analysis similar to the above[2] shows that this corresponds to a distance $[2\hbar R/(2meV)^{\frac{1}{2}}]^{\frac{1}{2}}$ at the emitter surface. If the two contributions to the minimum resolvable distance δ are combined vectorially,[5,20] the resultant resolution is

$$\delta = \left[\frac{2\hbar\beta R}{(2meV)^{\frac{1}{2}}} + \frac{16\beta^2 R^2 E_t}{eV} \right]^{\frac{1}{2}} \tag{2.18}$$

The energy E_t is of the order of $\frac{1}{4}(3kT + \alpha F^2)$ for an atom ionized before loss of energy to the emitter surface and, from the preceding discussion, about kT for an atom ionized after full accommodation. The fact that $\frac{1}{2}\alpha F^2/kT$ lies between 10 and 100 in practice indicates the importance of the accommodation process in determining the resolution of the low-temperature field-ion microscope. For helium ionization at 21°K, at a field of 4.5 V/Å, equation (2.18) predicts a resolution of $\frac{1}{2}$Å for an emitter of 100-Å radius, of which $\frac{1}{3}$ Å is due to the diffraction effect, although for emitters of larger radius and for heavier gases the diffraction contribution is unimportant. The predicted resolution for an emitter of 1000-Å radius under the same conditions is about 1 Å. Since the observed resolution of the field-ion microscope is between 2 and 3 Å at best, it is clear from the above discussion that the modulation of the electric field strength above the emitter surface is the limiting factor, although the effect of the finite radius of the imaging gas atom has yet to be investigated theoretically.

References

1. R. Stratton, *Phys. Rev.* **125**: 67 (1962).
2. R. Gomer, *Field Emission and Field Ionization*, Harvard University Press (Cambridge, Mass.), 1961.
3. R. H. Fowler and L. W. Nordheim, *Proc. Roy. Soc. (London)* **A119**: 173 (1928).
4. W. P. Dyke and W. W. Dolan, *Advan. Electron. Electron Phys.* **8**: 89 (1956); also E. E. Martin, J. K. Trolan, and W. P. Dyke, *J. Appl. Phys.* **31**: 782 (1960).
5. R. H. Good and E. W. Müller, *Handbuch der Physik* Vol. 21, Springer (Berlin), 1961, p. 176.
6. R. D. Young, *Phys. Rev.* **113**: 110 (1959).
7. R. D. Young and E. W. Müller, *Phys. Rev.* **113**: 115 (1959).
8. A. Van Oostrom, Doctorate Thesis, University of Amsterdam, 1965.
9. R. W. Wood, *Phys. Rev.* **5**: 1 (1897).
10. E. W. Müller, *Z. Physik* **106**: 541 (1937).
11. E. W. Müller, *Z. Physik* **108**: 668 (1938).
12. C. Herring and M. H. Nichols, *Rev. Mod. Phys.* **21**: 185 (1949).
13. C. Herring, *Metal Interfaces*, ASM Seminar Vol. 1951, p. 1.
14. J. H. Juretschke, in: *The Surface Chemistry of Metals and Semiconductors*, H. C. Gatos, ed., Wiley (New York), 1960, p. 38.
15. R. Smoluchowski, *Phys. Rev.* **60**: 661 (1941).
16. J. A. Becker, *Solid State Phys.* **7**: 379 (1958).
17. G. Ehrlich, *Proc. 3rd Intern. Congr. Catalysis*, 1964, W. M. H. Sachtler, G. C. A. Schuit, and P. Zweitering, eds., North–Holland (Amsterdam), 1965, p. 113.
18. R. D. Young and E. W. Müller, *J. Appl. Phys.* **33**: 91 (1962).
19. L. Schmidt and R. Gomer, *J. Chem. Phys.* **42**: 3573 (1965).
20. E. W. Müller, *Advan. Electron. Electron Phys.* **13**: 83 (1960).
21. E. W. Müller, *Z. Physik* **136**: 131 (1951).
22. M. G. Inghram and R. Gomer, *J. Chem. Phys.* **22**: 1279 (1954).
23. E. W. Müller, *J. Appl. Phys.* **27**: 474 (1956).

24. E. W. Müller and K. Bahadur, *Phys. Rev.* **102**: 624 (1956).
25. M. J. Southon, Ph.D. Thesis, University of Cambridge, 1963.
26. T. T. Tsong and E. W. Müller, *J. Chem. Phys.* **41**: 3279 (1964).
27. T. T. Tsong and E. W. Müller, *12th Field-Emission Symposium*, Pennsylvania State University, University Park, Pa., 1965.
28. D. S. Boudreaux and P. H. Cutler, *12th Field-Emission Symposium*, Pennsylvania State University, University Park, Pa., 1965.
29. M. J. Southon and D. G. Brandon, *Phil. Mag.* **8**: 579 (1963).
30. E. W. Müller, *J. Appl. Phys.* **28**: 1 (1957).
31. E. W. Müller and R. D. Young, *J. Appl. Phys.* **32**: 2425 (1961).

Chapter 3

FIELD EVAPORATION

D. G. Brandon†

3.1   Introduction

By applying an electric field to the tip of a field-ion microscope specimen it is possible to evaporate protruding surface atoms as ions. This process, known as *field evaporation*,[1] is thermally activated and has an activation energy dependent on the applied field strength. At a sufficiently high field strength the activation energy can be reduced to an arbitrarily low value, and so field evaporation can be made to take place at any temperature.

Field enhancement over protruding surface atoms reduces the activation energy for field evaporation in these regions, so that atomic protrusions at the surface of a field-ion microscope tip are evaporated preferentially.[2] Field evaporation is therefore used to clean and smooth the tip before examination of the field-ion image and thus represents the final stage of specimen preparation. By continued field evaporation successive lattice planes can be removed from the surface and the structure of the bulk lattice revealed. If the evaporation field is carefully controlled, evaporation can be limited to just a few atoms at a time and a detailed reconstruction of the crystal lattice is then possible, at least in principle.[3,4] Field evaporation is thus a most important tool not only for the final preparation of the specimen but also for the subsequent study of the lattice structure. Indeed it is only when a series of field-ion images are available in a single field-evaporation sequence that reasonably exact image interpretation becomes possible.

The term *field evaporation* is usually taken to refer to the evaporation of a metal from its own lattice, and the term *field desorption* is used to describe the analogous process for foreign atoms at the surface. Field evaporation could therefore be described as a special case of field desorption. However the terms *adsorption* and *desorption* usually refer to gaseous

† Senior Research Scientist. Battelle Memorial Institute. Geneva. Switzerland. Presently Professor of Metallurgy. The Technion. Israel Institute of Technology. Haifa. Israel.

contaminants at a surface, while *condensation* and *evaporation* are the analogous terms for components which are liquid or solid in their standard state. This difference in terminology relates to no fundamental physical difference in the processes involved, and so we will use the term field evaporation in a more general sense to include evaporation of impurities and solute atoms as well as evaporation of the solvent matrix.

The first experimental work on field evaporation preceded the development of the field-ion microscope and the early work, on barium and thorium adsorbed on tungsten[1] (hence, field desorption), is still of considerable importance. Later experimental work was limited to adsorbates with binding energies considerably less than the solvent matrix,[5-8] and a detailed theory of field desorption has been developed for this case.[9,10] The field evaporation of the matrix has been considered semiquantitatively,[2,11] and experimental data have for the most part confirmed this semiquantitative picture.[12] More recently, the field evaporation of dilute alloys and impurity-containing solutes has been reexamined[13] and the semiquantitative picture made rather more quantitative.

3.2   Field-Evaporation Models

The potential energy of an atom or ion near a metal surface can be represented to a first approximation by a single curve.[9] At a given distance from the surface the potential energy is assumed to be single valued, so that energy changes only occur normal to the surface and thermally activated processes can be described in terms of a one-dimensional energy barrier. Clearly such a model can only be strictly valid at distances which are large compared with the dimensions of the surface structure; however, at any fixed distance from the metal, minima will exist in the potential energy surface corresponding to this distance, and by drawing the potential energy curve to follow these minima the validity of the one-dimensional model can be extended into small distances *as long as the minima lie in a straight line.* This last is important, since otherwise a particle would require translational energy in at least *two* directions to be activated up the curve.[14] Fortunately the dimensions of the activation barrier in field evaporation are small compared to any possible curvature of the surface-energy "valleys," and so one-dimensional potential-energy curves provide a suitable framework for discussing field evaporation, even though the details of the energy barrier are on the same scale as the structure of the surface.

The potential energy of an atom, $V_A(x)$, as a function of distance x from the surface is shown in Fig. 3.1a. The minimum corresponds to the sublimation energy of the atom Λ. At smaller distances than the minimum the repulsion of the atom by its neighbors raises the potential energy, and at larger distances the attractive forces are smaller. A similar curve for an

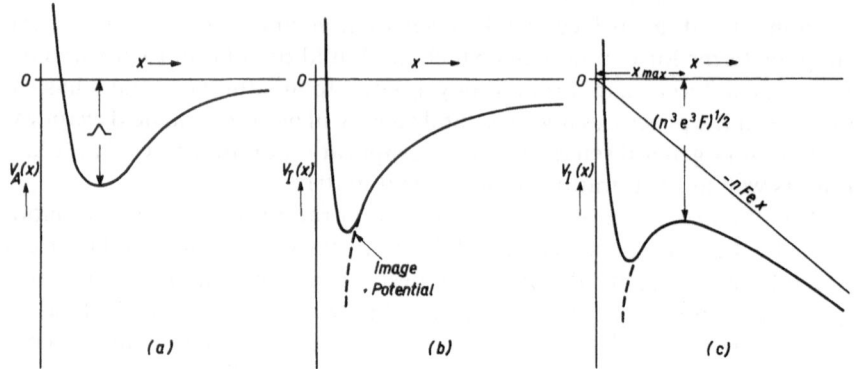

Fig. 3.1. Potential energy curves (a) for an atom, (b) for an ion, and (c) for an ion in the presence of a field.

ion is shown in Fig. 3.1b. The energy required to ionize an atom in the gas phase is $\sum_n I_n$, where n is the charge on the ion and I_n is the nth ionization potential of the atom. If the electrons are returned to the metal, an energy $n\phi_0$ is recovered, where ϕ_0 is the absolute work function of the metal and corresponds to the energy difference between an electron at *infinity* and an electron at the top of the Fermi surface in a finite crystal of the metal. The zero level for the potential energy of an ion, $V_I(x)$, must therefore lie at a distance $\sum_n I_n - n\phi_0$ above the zero level for the corresponding atom. As the ion approaches the surface, it will be attracted by its negative image in the metal (the image potential), but at small distances the repulsion potential due to the interaction between the ion and the surface atoms will become strong and a minimum in the potential curve will appear (Fig. 3.1b). On applying an electric field to the specimen the potential energy of the ion will be reduced by a factor $nFex$, where e is the charge on an electron and F is the applied field, and a maximum will appear in the ion energy (Fig. 3.1c). At distances which are sufficiently large to be able to ignore the repulsion potential, the potential energy of the ion is given by

$$- V_I(x) = nFex + \frac{n^2 e^2}{4x} \tag{3.1}$$

Differentiation to obtain the energy maximum yields

$$\frac{1}{x_{max}} = \sqrt{\frac{4F}{ne}} \quad \text{and} \quad - V_I(x_{max}) = (n^3 e^3 F)^{\frac{1}{4}} \tag{3.2}$$

The energy at this maximum, referred to as the *image* or *Schottky hump*,

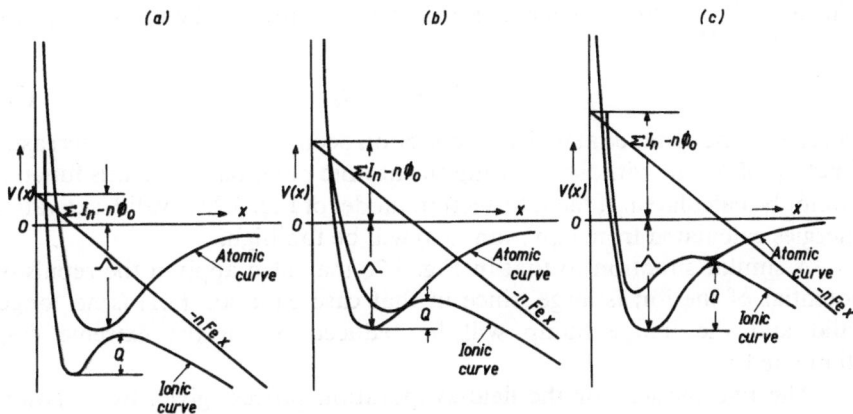

Fig. 3.2. Superposition of potential energy curves (schematic) for (a) ionic state stable and (b) transition from atomic to ionic state, followed by evaporation over an image-potential hump, with (c) potential maximum at transition from atomic to ionic states.

is independent of x as long as the neglect of the repulsion potential implied by equation (3.1) is justified.

If the ionic potential curve of Fig. 3.1c is superimposed on the atomic potential curve of Fig. 3.1a, several possible situations may occur,[9,10] which are illustrated in Fig. 3.2. Firstly, if the ionization potential is sufficiently low, the atom may be adsorbed at the surface as an ion (cesium or barium adsorbed on tungsten are good examples of this) so that the ionic potential curve lies below the atomic potential curve and field evaporation occurs over the image-potential hump described by equation (3.2) and shown in Fig. 3.2a. More common is the condition shown in Fig. 3.2b, where the atomic state is stable but the ionic and atomic curves cross over to the left of the image hump. The minimum in the atomic curve lies at a distance $\Lambda + \sum_n I_n - n\phi_0$ below the zero level of equations (3.1) and (3.2) (cf. Figs. 3.1a and b), so that the activation energy for field evaporation Q is given by[1]

$$Q = Q_0 - (n^3 e^3 F)^{\frac{1}{2}} \qquad (3.3)$$

where[10]

$$Q_0 = \Lambda + \sum_n I_n - n\phi_0$$

At higher ionization potentials the image-potential hump may lie above the atomic-potential curve. In that case the energy maximum does not correspond to this hump but to the intersection of the two curves

(Fig. 3.2c). The activation energy for field evaporation is then given approximately by[10,15]

$$Q = Q_0 - nFex_c \tag{3.4}$$

where x_c is the distance at which intersection occurs. Clearly x_c is a decreasing function of F, but without knowing the atomic potential curve this function cannot be calculated. If the intersection model of Fig. 3.2c is valid, activation energies calculated from equation (3.3) will be too high.

A similar situation to that in Fig. 3.2c may also apply if the repulsion potential of the ion is large, since in that case equation (3.1) is no longer valid and the image hump will be reduced or, in the extreme case, eliminated.[15]

The two models for the field-evaporation process given by equations (3.3) and (3.4) will be referred to as the *image potential* and *intersection* models, respectively. The transition from one model to the other obviously depends to some extent on the applied field strength, but for the cases of interest, where $Q \ll Q_0$, the transition depends principally on the ratio

$$\chi = \left(\sum_n I_n - n\phi_0 \right) \bigg/ \Lambda \tag{3.5}$$

For $\chi < 1$ Fig. 3.2a should apply, for $1 < \chi < 3$ Fig. 3.2b is likely to be valid, and for $\chi > 3$ Fig. 3.2c almost certainly represents the true situation.

3.3   Calculation of Evaporation Fields

For metals the image-potential model of Fig. 3.2b provides a reasonable first approximation and values for the activation energy for field evaporation can be calculated from equation (3.3). At low temperatures Q is small, so that the evaporation field is determined principally by Q_0.

$$(e^3 F)^{\frac{1}{2}} \cong n^{-\frac{3}{2}} Q_0 = Q'_n \tag{3.6}$$

The condition that Q'_n should be a minimum determines the charge on the evaporating ion, and this value of n then determines the evaporation field F_e.[11] For pure metals Q'_n can be calculated from the known values of the sublimation energy and ionization potentials. Instead of the absolute work function ϕ_0 it is usual to take the measured work function ϕ, when calculating Q'_n, since this includes a correction for the double-layer potential (see below). Values of Q'_n for pure metals are given in Table 1, together with the calculated evaporation fields and expected ionic species.[11,12] From Table 1 it is seen that most metals are expected to evaporate as doubly charged ions.

In order to calculate Q'_n for solute atoms and impurities, we need to know the sublimation energy of the evaporating species. If the partial molar

TABLE 1

Calculated Evaporation Fields

Z	Element	Q'_1, eV	Q'_2, eV	F_e V/Å	Ion Expected
4	Be	8.87	8.15	4.6	Be^{2+}
5	B	9.96	11.6	6.5	B^+
6	C	14.3	10.9	8.2	C^{2+}
12	Mg	5.46	5.85	2.4	Mg^{2+}
13	Al	5.07	6.96	1.8	Al^+
14	Si	8.25	7.0	3.4	Si^{2+}
20	Ca	4.90	4.80	1.6	Ca^{2+}
22	Ti	7.51	5.8	2.3	Ti^{2+}
23	V	7.64	6.1	2.6	V^{2+}
24	Cr	6.74	6.68	3.1	Cr^{2+}
25	Mn	6.30	6.35	2.8	Mn^{2+}
26	Fe	8.00	7.19	3.6	Fe^{2+}
27	Co	7.86	7.35	3.7	Co^{2+}
28	Ni	7.08	7.15	3.5	Ni^{2+}
29	Cu	6.67	7.90	3.1	Cu^+
30	Zn	6.44	7.10	3.5	Zn^{2+}
31	Ga	4.74	7.2	1.6	Ga^+
32	Ge	7.05	6.50	2.9	Ge^{2+}
40	Zr	9.05	6.67	3.1	Zr^{2+}
41	Nb	9.74	7.0	3.4	Nb^{2+}
42	Mo	8.98	7.4	3.8	Mo^{2+}
44	Ru	8.36	7.5	3.9	Ru^{2+}
45	Rh	8.43	7.4	3.8	Rh^{2+}
46	Pd	7.42	7.7	4.1	Pd^{2+}
47	Ag	5.78	7.96	2.3	Ag^+
48	Cd	6.12	6.70	3.1	Cd^{2+}
50	Sn	6.08	5.75	2.3	Sn^{2+}
57	La	6.64	5.50	2.5	La^{2+}
73	Ta	11.7	8.4	4.6	Ta^{2+}
74	W	12.1	9.0	5.7	W^{2+}
75	Re	11.1	7.9	4.3	Re^{2+}
76	Os	11.1	8.4	4.6	Os^{2+}
77	Ir	10.7	8.0	4.4	Ir^{2+}
78	Pt	9.26	7.7	4.1	Pt^{2+}
79	Au	8.10	8.20	4.3	Au^{2+}
80	Hg	6.51	7.35	2.9	Hg^+
81	Tl	3.94	7.14	1.1	Tl^+
82	Pb	5.41	5.78	2.3	Pb^{2+}
83	Bi	6.6	6.2	2.6	Bi^{2+}

enthalpy of the component in dilute solution, $\overline{\Delta H_0}$, or the heat of solution, H_s, is known, the sublimation energy is approximately given by[13]

$$\Lambda = \Lambda_0 - \overline{\Delta H_0} \qquad (3.7a)$$

or

$$\Lambda = \Lambda_0 + H_s \qquad (3.7b)$$

where Λ_0 is the sublimation energy of the pure component.

This equation is not quite correct because it does not take into account differences between the binding energy of the solute at the surface and in the matrix (segregation effects), and the value of Λ given by equations (3.7) is in general too low. When heats of adsorption H_a have been measured,[16]

TABLE 2
Heat of Solution, Adsorption, and Formation

System		H_s,[18] kcal/g atom of solute	H_a,[16] kcal/g atom of gas	$-\Delta H_F$,[17] kcal/g mol of compound	Compound
$H_2 +$	Mo	−6.0	20		
	Ni	−5.1	12		
	Pt	−17.4	13		
	Pd	+0.86	13		
	Cr	−5.7	22		
	Fe	−6.5	16		
	Cu	−13.9	....		
$O_2 +$	Si	94.0	115	105	$\frac{1}{2}SiO_2$
	W	....	97	67	$\frac{1}{2}WO_2$
	Mo	63.4†	86	70	$\frac{1}{2}MoO_2$
	Ni	10.5	57.5	57.5	NiO
	Pt	....	33.5	10.2	$\frac{1}{4}Pt_3O_4$
	Cr	54.0	....	90	$\frac{1}{3}Cr_2O_3$
	Fe	37.0	....	63.2	FeO
$N_2 +$	W	....	42.5	17.2	W_2N
	Mo	−2.6†	51.6	16.6	Mo_2N
	Ni	....	5	−0.2	Ni_3N
	Cr	+1.8†	....	25.2	Cr_2N
	Fe	−14.1	....	2.6	Fe_4N
$C +$	W	−17.6†		9.1	WC
	Cr	−10.6†		16.4	Cr_4C
	Fe	−19.9[19]		−5.4	Fe_3C
	Mn	−2.0†		3.6	Mn_3C
	Co	−10.2†		−9.3	Co_3C

†Values based on the reported solid solubility and the known equilibrium diagram.

they are generally found to be greater than the heat of formation of the corresponding compound,[17] $-\Delta H_F$. The heat of solution H_s, on the other hand,[18,19] is generally less than $-\Delta H_F$ (Table 2). The difference $H_o - H_s$ appears as a reduction in the binding energy of neighboring solvent atoms at the surface, so that preferential evaporation of solvent atoms may indicate an underlying impurity atom.

For gaseous impurities the dissociation energy E_d must be substituted for the sublimation energy Λ_0, so that we have

$$\Lambda = E_d + H_a \tag{3.8}$$

for such components.

Values of Λ have been calculated for a number of solutes in iron and the corresponding values of Q'_n evaluated.[13] The results are given in Table 3. The high values of Q'_n determined for the nonmetallic solutes are a consequence of the high ionization potentials of these elements. This is further brought out by Table 4, which gives similar results for oxygen, nitrogen, and carbon dissolved in various metals. For such components the intersection model for field evaporation (Fig. 3.2c) is probably correct, but an alternative possibility, evaporation as a molecular ion, must also be considered.

The evaporation of a molecular ion can be treated in exactly the same way as the evaporation of a monatomic ion. The comparison of the energy

TABLE 3

Field Evaporation of Solute Atoms in Iron
$\phi = 4.17$ eV

	$-\overline{\Delta H_0}$,[11] or H_s,†	Λ‡	$\sum_n I_n$, eV		Q'_n, eV	
	eV	eV	1	2	1	2
Fe	....	4.20	7.90	24.4	8.00	7.19
Cr	−0.13	4.00	6.74	23.4	6.57	6.75
Mn	0.05	2.77	7.43	23.1	6.03	6.23
Co	0.00	4.40	7.86	25.5	8.09	7.64
Ni	0.07	4.43	7.63	25.8	7.89	7.78
Cu	−0.35	3.05	7.72	27.9	6.60	8.03
Al	0.44	3.74	5.97	24.8	5.54	7.17
Si	1.26	6.16	8.15	24.5	10.14	7.90
C	−0.87	6.60	11.22	35.49	13.60	11.90
N	−0.62	3.10	14.48	43.95	13.40	13.70

†Table 2.
‡Values based on available data for H_s or $-\overline{\Delta H_0}$ and on known values of Λ_0 and E_d.[17]

TABLE 4
Energy to Field-Evaporate Nonmetallic Impurities

	Λ, eV†				φ, eV	$\sum_n I_m$, eV		Q'_1, eV				Q'_2, eV			
	Metal	O	N	C		1	2	Metal	O	N	C	Metal	O	N	C
Fe	4.2	4.2	3.1	6.6	4.17	7.9	24.4	8.00	13.6	13.4	13.6	7.19	15.7	13.7	11.9
Ni	4.4	3.0	3.7	7.0	5.01	7.63	25.8	7.08	11.5	13.2	13.2	7.15	14.7	13.3	11.5
Mo	6.8	5.4	6.0	7.3	4.30	7.13	23.3	8.98	14.6	16.2	14.2	7.4	16.0	14.6	12.7
W	8.7	5.5	4.5	6.7	4.52	7.98	25.7	12.10	14.6	14.5	13.4	9.0	15.9	14.0	11.8
Pt	5.3	3.0	...	...	5.32	8.96	27.5	9.26	11.2	...	...	7.7	14.5		
$\sum_n I_g$, eV 1		13.55	14.48	11.22											
2		48.48	43.95	35.49											

†Values based on available data for H_a, H_s, or ΔH_F.

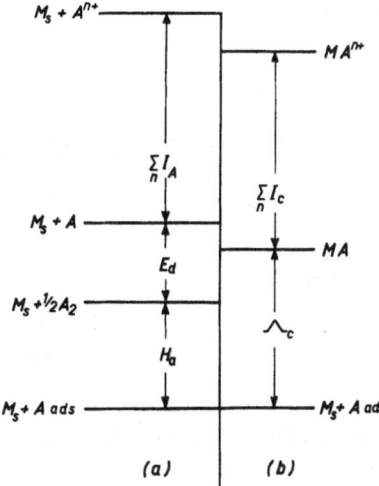

Fig. 3.3. Energy schemes for evaporation of (a) a nonmetallic impurity and (b) the corresponding molecular ion.

schemes applicable to these two processes is given in Fig. 3.3, and for a gaseous adsorbate the condition that the molecular ion should have the lower value of Q'_n is

$$E_d + H_a - \Lambda_c > \sum_n I_c - \sum_n I_a \tag{3.9}$$

where Λ_c is the sublimation energy of the molecular complex and I_c and I_a refer to the ionization potentials of the molecular complex and the adsorbate respectively. Values for I_c and Λ_c are not available, but if H_a is large and the molecular species are reasonably volatile, equation (3.9) will probably be fulfilled. Mulson and Müller[20] have postulated the existence of W—O and W—N molecular complexes to explain some of their observations on the field evaporation of these adsorbates.

3.4 Experimental Determination of the Field-Evaporation Parameters

If field evaporation is a thermally activated process, the evaporation rate k_e will be given by

$$k_e = v \exp - Q/kT \tag{3.10}$$

where the frequency factor v is to a first approximation temperature independent.

By taking logarithms and substituting from equation (3.3)

$$-kT \ln k_e/v = Q_0 - (n^3 e^3 F)^{\frac{1}{2}} \tag{3.11}$$

Hence

$$\left(\frac{\partial \ln k_e}{\partial F}\right) = \frac{-\frac{1}{2}(n^3 e^3 F)^{\frac{1}{4}}}{F(kT)} \qquad (3.12)$$

or, since $dF/F = d \ln F$ and $Q \ll Q_0$,

$$\left(\frac{\partial \ln k_e}{\partial \ln F}\right)_T = \frac{1}{2}\frac{Q_0}{kT} \qquad (3.13)$$

Similarly

$$\left(\frac{\partial(kT)}{\partial \ln F}\right)_{k_e} = -\frac{1}{2}\frac{Q_0}{\ln(k_e/v)} \qquad (3.14)$$

From an experimental determination of both equations (3.13) and (3.14) it is in principle possible to measure Q_0 and v. Some results for tungsten, molybdenum, and platinum are summarized in Table 5.[12]

If the evaporation field is measured and the ionic charge is known, Q_0 can also be determined directly from equation (3.3). The average electric field strength at the tip depends on the tip size R and the electrode geometry and is given by an equation of the form[21]

$$F = b\frac{V}{R} \qquad (3.15)$$

where V is the applied voltage and b is a semiconstant

$$\frac{1}{b} \cong \frac{1}{2}\ln\frac{D}{R} \qquad (3.16)$$

D being a characteristic length of the order of the tip-to-cathode distance in the microscope.

The tip size R can be determined from electron micrographs of tip profiles[22] or by averaging the local tip radii derived from field-ion micrographs.[23] Neither method is very satisfactory, and field strengths determined

TABLE 5
Field Evaporation Parameter

| | $\left(\dfrac{\partial \ln(kT)}{\partial \ln F}\right)_{k_e}$ $k_e = 0.1 \ \text{sec}^{-1}$ | $\left(\dfrac{\partial \ln k_e}{\partial \ln F}\right)_T$ $T = 88°K$ | Q_0, eV | | | v sec^{-1} |
			From rate measurements	From evaporation	Calculated	
W	−0.075	267	4.1	27.8	25.4	5×10^{10}
Mo	−0.13	275	2.8	23.9	20.9	10^6
Pt	−0.072	169	2.6	21.7	21.7	7×10^6

from equations (3.15) and (3.16) are generally not accurate to better than 20%. If the average work function of the metal is known, the ratio of b/R in equation (3.15) can be determined directly from a Fowler–Nordheim plot.[24] Unfortunately this is only possible if an ultrahigh vacuum system is available. Such experimental measurements as are available indicate that the evaporation field for a small specimen rises rapidly until the stable field-evaporation end form is reached and then remains constant.[25,26] By assuming equation (3.3) to hold, values of Q_0 for doubly charged ions have been determined from the measured evaporation fields of molybdenum, platinum, and tungsten at liquid-nitrogen temperature, and these are compared in Table 5 with the theoretical values of Q_0 (Table 1) and the measured values determined from the field sensitivity of the evaporation rate (equation (3.13)]. Clearly the field sensitivity of the evaporation rate predicted from equations (3.3) and (3.10) is seriously in error and we shall see below that this might be explained by surface polarization effects.

By combining equations (3.13) and (3.14) the frequency factor v can be determined independently of Q_0. From Table 5 we see that $v \ll 10^{13} \sec^{-1}$, which is the frequency factor expected for a thermally activated process in a solid.

This is probably a result of partial tunneling of the metal ion through the potential barrier.* The lower effective value of Q is then paid for by a reduced frequency factor. At the low temperatures at which field evaporation is normally performed ($< 100°K$), large changes in the frequency factor have very little effect on the basic process. Thus a change of four orders of magnitude in v is equivalent [equation (3.11)] to a change in Q of only $10 kT$.

3.5 Surface Electronic Structure and Polarization Corrections

The energy required to extract an electron from a metal surface is not equal to the *absolute* work function ϕ_0 because the electron is not taken to infinity but to some point outside the surface at a distance that is small compared with the specimen dimensions. This energy, the average work function ϕ, is dependent on the electronic structure of the surface and hence on the surface orientation. The structure dependence arises from the presence of an electrical double layer that sets up a difference in the potential seen by an electron just inside and just outside the surface. Two different effects are possible[27]: A spreading of the conduction electron cloud out beyond the positive ion cores gives rise to a double layer with the negative side outward, which increases the work function (Fig. 3.4a); while a draining away of the conduction electron cloud between the ion cores gives rise to a double layer of opposite sign, which tends to reduce the work function

* Tsong (to be published) has recently proposed a more likely explanation.

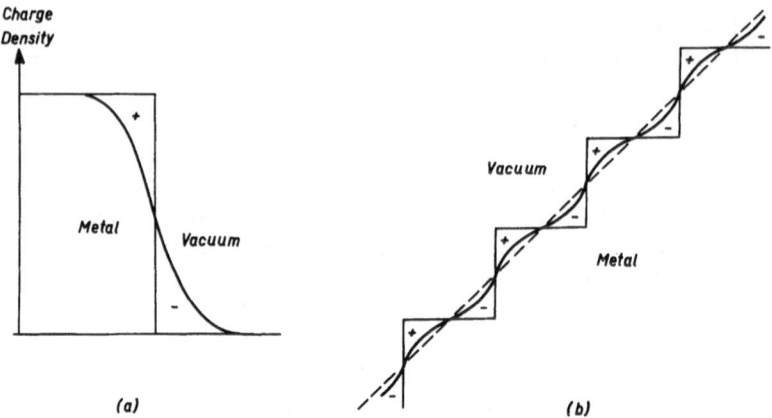

Fig. 3.4. Surface double layer due to (a) electron spreading and (b) electron smoothing.

(Fig. 3.4b). Calculations indicate that the latter effect is the more important, and it is consistently observed that the lowest work function of a metal is obtained for the high-index loosely packed planes.

The effect of the double-layer potential on the potential-energy curves given in Figs. 3.1 and 3.2 is to raise the ionic potential curves by an amount equal to n times the double-layer potential, and in calculating values of Q'_n we have already taken account of an "average" double-layer potential by substituting the measured average work function ϕ for the absolute work function ϕ_0 in equation (3.3).

Because of the double layer of electric charge the field at the site of the evaporating ion will be partially screened. The distance involved should be of the order of the screening distance in the bulk metal,[28] typically about 0.5 Å. The effect is equivalent to a reduction in the potential energy of the surface atoms that arises from a permanent dipole moment.

The applied field strength at the surface is balanced by an accumulation of surface charge which is obtained by displacing the conduction electron cloud into the metal and exposing the positive ion cores. The effect may be regarded either as a field dependence of the field-penetration distance or as a result of the polarizability of the electrical double layer. On a loosely packed plane the charge carried by the surface atoms must be greater at a given field strength than on a closely packed plane, so that the effective polarizability should be smallest for the close-packed planes.

The three correction factors can be written as

$$\Delta Q_0 = A + BF + CF^2 \qquad (3.17)$$

corresponding to the effect of a double-layer potential $V_d = A/ne$, a dipole moment $\mu = B \cong ne/q$, and a polarizability $\alpha = 2C$. The double-layer potential varies from approximately zero for the close-packed planes to about 2 V for the high-index planes. For the transition metals q^{-1} is about 0.3 Å, so that, at an evaporation field of 4 V/Å, $BF \cong 2.5$ eV. The surface polarizability is expected to be of the order of 1 to 10 Å, depending on the density of packing of the plane considered, so that the polarizability correction should be of the order of 0.5 to 5 eV at a field strength of 4 V/Å. Ignoring the double-layer potential, which has already been accounted for in the values of Q'_n given in Table 1, the total polarization correction factor for Q_0 should be of the order of 5 eV, varying with orientation. Comparison with Table 1 suggests that $\Delta Q_0/Q_0 \cong 0.2$ with a corresponding increase in the evaporation field of about 10% if the image-force theory is correct. The effect of the polarization corrections on the evaporation field calculated from equation (3.3) is therefore not very serious.

The effect of polarization on the field and temperature sensitivity of the evaporation rate should be much larger. Including ΔQ_0 in equation (3.3) and differentiating yields

$$F \left(\frac{\partial Q}{\partial F} \right)_T = \tfrac{1}{2}(n^3 e^3 F)^{\frac{1}{2}} + BF + 2CF^2 \tag{3.18}$$

So that, by combining with equation (3.10) and remembering $Q \ll Q_0$,

$$\left(\frac{\partial \ln k_e}{\partial \ln F} \right)_T = \frac{1}{2} \frac{Q_0}{kT} \left(1 - 2\frac{BF + 2CF^2}{Q_0} \right) \tag{3.19}$$

The factor on the right reduces to $[1 - 4(\Delta Q_0/Q_0)]$ if the CF^2 term predominates and to $[1 - 3(\Delta Q_0/Q_0)]$ if the two terms are about equal. From the results given in Table 5 we can calculate the approximate size of the polarization correction by combining equations (3.19) and (3.13) and obtaining

$$a \cdot \Delta Q_0 = Q_{0\,calc} - Q_{0\,exp} \tag{3.20}$$

where $Q_{0\,exp}$ is the value determined from equation (3.13), $Q_{0\,calc}$ is the value used in equation (3.3), and a is a constant

$$2 < a < 4$$

Setting $a = 3$, we find from the results given in Table 5 that $\Delta Q_0 \cong 0.28\,Q_0$ for tungsten. If the polarizability term is dominant, then $a = 4$, and by setting $\Delta Q_0 = \tfrac{1}{2}\alpha F_e^2$ we can obtain a value for the effective polarizability. From the results in Table 5 we then find $\alpha \cong 3.4$ Å³ for tungsten.

Obviously polarization effects could account for the discrepancy between the calculated and observed field sensitivity of the evaporation rate. Since equation (3.14) is derived directly from equation (3.11), precisely the same correction factor must be applied to the temperature sensitivity of the evaporation field.

3.6   Structure of Field-Evaporated Surfaces—Pure Metals[11]

With the discussion of polarization corrections we have introduced terms into the field-evaporation equations which depend on the site of the evaporating atom at the surface and can thus play a part in controlling the surface structure and endform of the field-evaporated specimen tip. The evaporation rate will be enhanced in any region where the average value of Q is low, and the subsequent flattening of the surface resulting from the higher evaporation rate will then reduce the local field strength in this region. The final end form thus corresponds to a situation where variations in the radius of curvature of the tip compensate for variations in the corrections to be applied to Q_0. Only parameters that show a systematic variation with orientation can affect this end form, i.e., the orientation dependence of the polarization and double-layer corrections and the variations in local field enhancement resulting from the surface geometry imposed by the lattice structure of the metal. If the applied field strength F_a is measured for each region of the surface separately (by field-emission experiments on the individual emitting areas), we can write the effective field strength F' as βF_a, where β is a local field-enhancement factor and depends on the lattice step height at the site of the evaporating atom and the distance from the atom to a hypothetical smooth section through the crystal lattice corresponding to the final end form (Fig. 3.5).

In a pure metal additional corrections to Q_0 arise from differences in binding energy of the surface atoms. This difference in binding energy, $\Delta\Lambda$, must be equal to the change in surface energy accompanying field

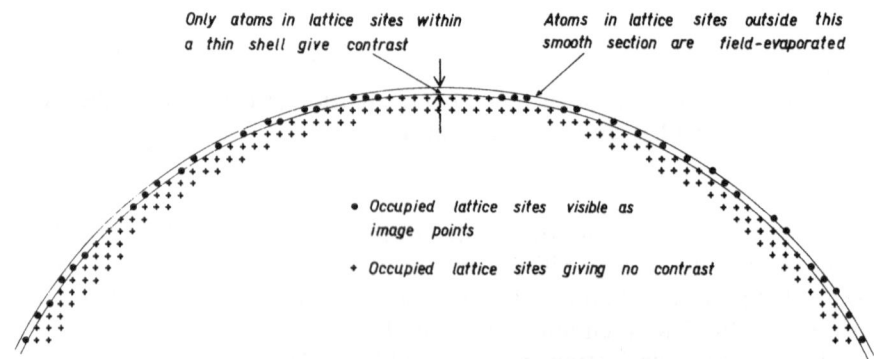

Fig. 3.5. Smooth section approximation to the field-evaporated surface.

evaporation of the atom and is small when evaporation proceeds from kink sites at lattice steps. Clearly this is not true for the evaporation of the the last few atoms on a close-packed plane, which in consequence tend to evaporate as a group.

Although changes in the local radius of curvature occur to balance differences in lattice step height and polarization effects, they do not result in a uniform value of Q at all evaporation sites. The density of suitable evaporation sites varies with surface orientation and is lowest in the region of the close-packed planes. Since the binding energy is approximately constant for all these sites, a stable end form can only be maintained if Q is somewhat lower in the regions of low site density. The specimen geometry imposes a further limitation since the evaporation rate per unit surface area must be a maximum at the tip apex and fall off toward the shank if the end form is to be stable (Fig. 3.6).

The most successful model for the field-evaporated surface and the one on which most image interpretation is based is the computer model developed by Moore.[29] The basic assumption is that the field-evaporated surface includes *all* the atoms whose sites lie within a smooth section taken through the crystal lattice and *no* atoms whose sites lie outside this section (Fig. 3.5). In order to obtain a computed field-ion image, Moore has made two further assumptions:

1. The atoms appearing in the image all lie within a fixed distance of the smooth section chosen.
2. This section is, to a good approximation, hemispherical.

The first of these secondary assumptions seems reasonable, but, as we have seen above, the second cannot be true. Even so, the correlation between the computed field-ion image and the actual image observed for

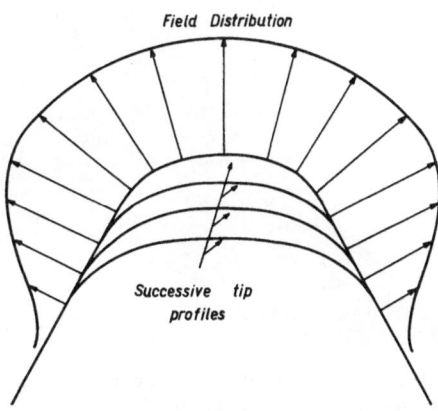

Fig. 3.6. Field distribution and evaporation rate over an isotropic tip.

pure metals is very satisfactory (Fig. 3.7),[29] and the smooth section approximation is generally used as a basis for image interpretation. Moore has
further suggested that for a perfect crystal the order of evaporation of the
atoms is in the order of their distance from the surface section. If a stable
end form is to be maintained, this last hypothesis must be statistically true.

If atoms are removed from inside the smooth section corresponding
to the stable end form or are retained outside this section, we can speak
of preferential field evaporation or retention respectively. Such effects give
rise to irregularities in the field-ion image and are easily confused with the
presence of defects in the crystal lattice. Most image interpretation depends

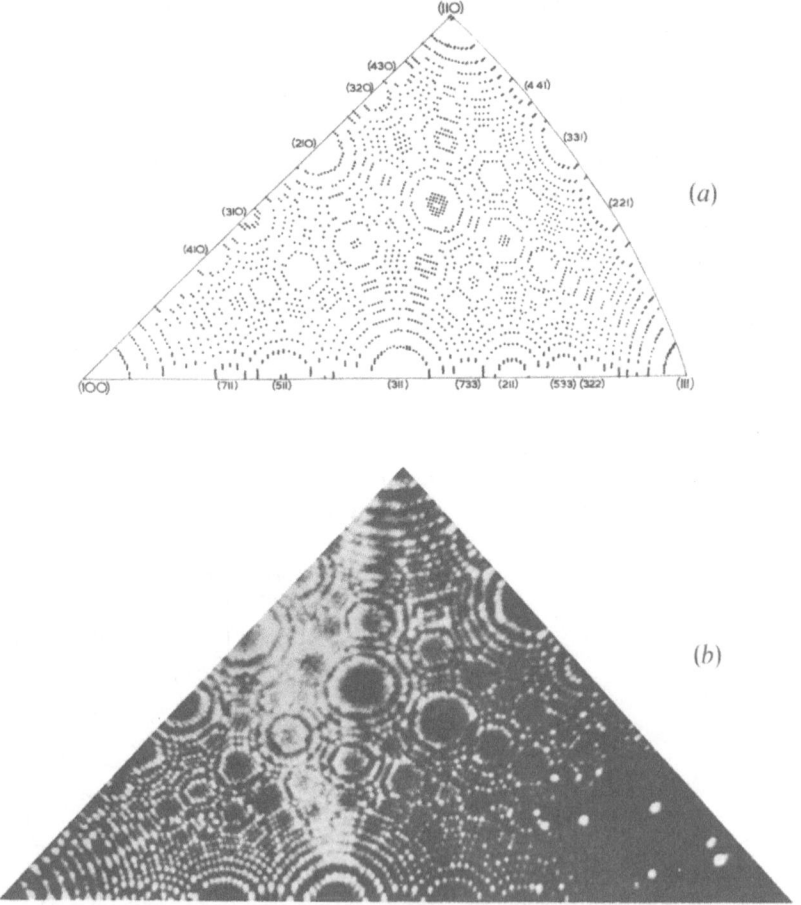

Fig. 3.7. Comparison between (a) a computed field-ion image and (b) the observed field-
ion image. (From A. J. W. Moore.[29])

on a correct distinction being made between irregularities in the crystal lattice and irregularities in the section taken through the lattice. Since the presence of lattice defects implies localized regions of high strain energy, lattice defects frequently give rise to preferential evaporation and corresponding surface irregularities so that image interpretation is seldom completely straightforward. Surface irregularities may also occur in the absence of lattice defects. Thus the increase in the effective binding energy of the surface atoms arising from polarization effects can sometimes stabilize atoms in suitable protruberant positions,[30] which gives rise to the *zone lines* commonly observed in many metals (Fig. 3.8).

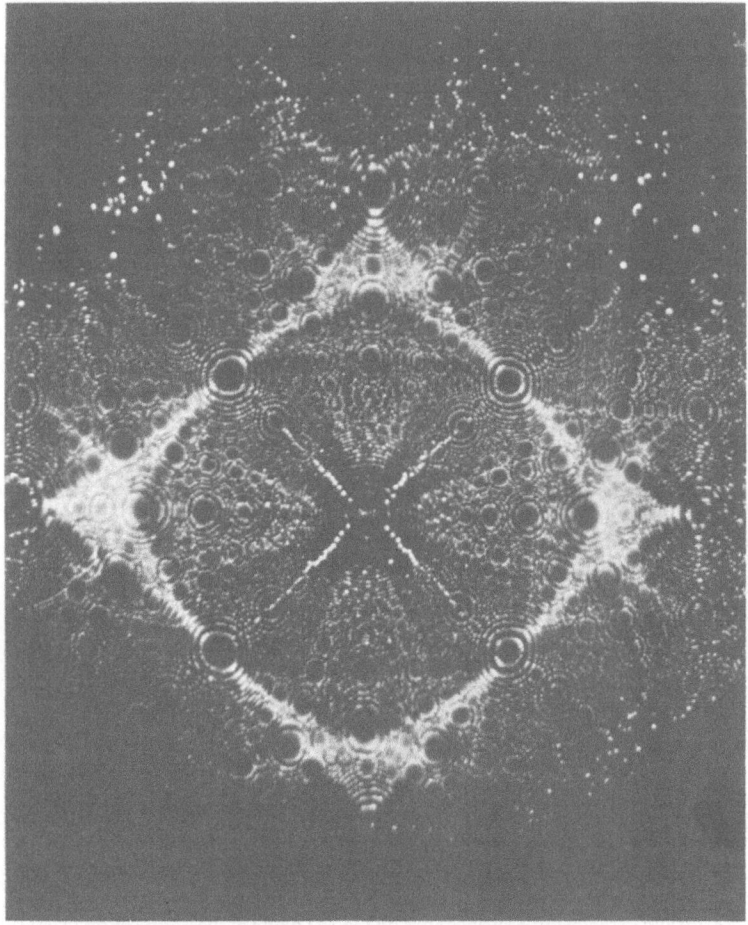

Fig. 3.8. Zone lines on platinum. (From E. W. Müller.[30])

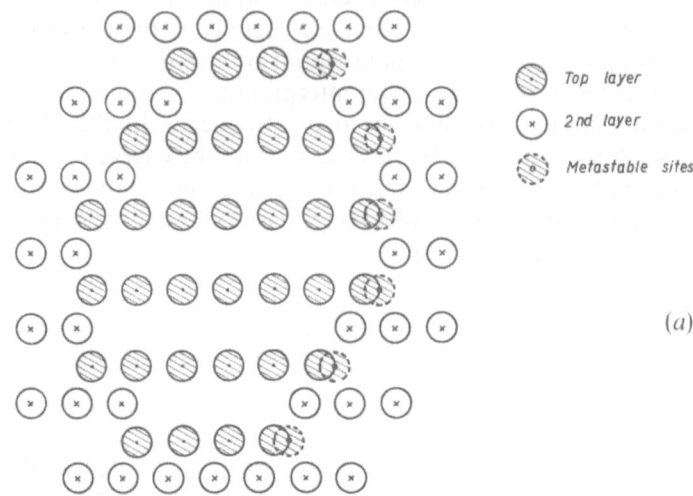

Top layer

2nd layer

Metastable sites

(a)

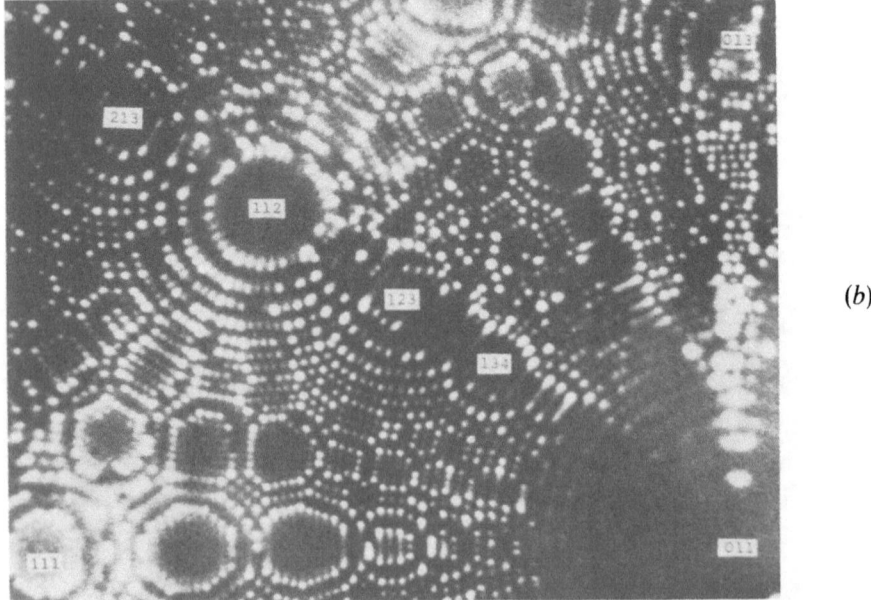

(b)

Fig. 3.9. (a) Origin of metastable surface sites on the 112 plane of the body-centered cubic lattice; (b) atoms in metastable sites on a tungsten tip. (From E. W. Müller.[30])

Even within the smooth section it is sometimes possible for atoms to occupy metastable sites at the surface which have no existence in the bulk lattice.[30] This is illustrated in Fig. 3.9. The metastable sites can give rise to apparent extra half-planes of atoms which could lead to a serious misinterpretation of the image, while single atoms in metastable sites can easily be mistaken for adsorbed impurities.

Lattice defects may have two distinct but related effects, firstly on the surface section and secondly on the order of evaporation. Vacant lattice sites on close-packed planes lower the binding energy of neighboring atoms and give rise to preferential removal of these atoms at the surface. Because preferential removal of one atom results in both local field enhancement and a reduction in the binding energy of neighboring atoms, the atoms around a vacant lattice site tend to evaporate as a group. On loosely packed lattice planes containing no nearest neighbor atoms single vacancies can be observed, but there is still some doubt as to the effect of a vacancy in underlying planes on the field evaporation sequence. Self-interstitial atoms are not stable at the surface and relax into normal lattice sites. It has been suggested that surface relaxation over underlying interstitial atoms (Fig. 3.10a) will give rise to field enhancement over a large region and at the same time reduce the image force tending to pull the interstitial out to the surface.[31] Surface relaxation will also reduce the energy barrier to interstitial migration, and it is uncertain to what extent surface relaxation is possible before the interstitial moves into a stable site either at the surface (Fig. 3.10b) or, more likely, by pushing a neighboring atom to the surface (Fig. 3.10c). Interstitial atoms relaxing to the surface will be field evaporated at a reduced field strength, and, unless they lie in protected positions, evaporation may occur before they are observed. Such protected positions are the centers of the close-packed planes (where the applied field strength is low) or the well-shielded lattice sites at the surface of loosely packed planes. Interstitial atoms appearing at step edges are even better protected but cannot readily be distinguished from the original atoms.

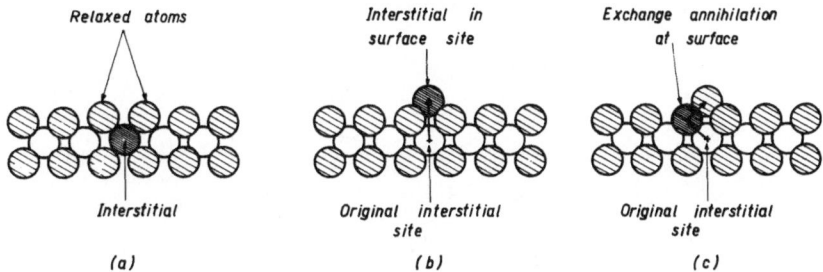

Fig. 3.10. (a) Relaxation over a self-interstitial atom, (b) migration of an interstitial to a surface site, and (c) exchange migration of an interstitial.

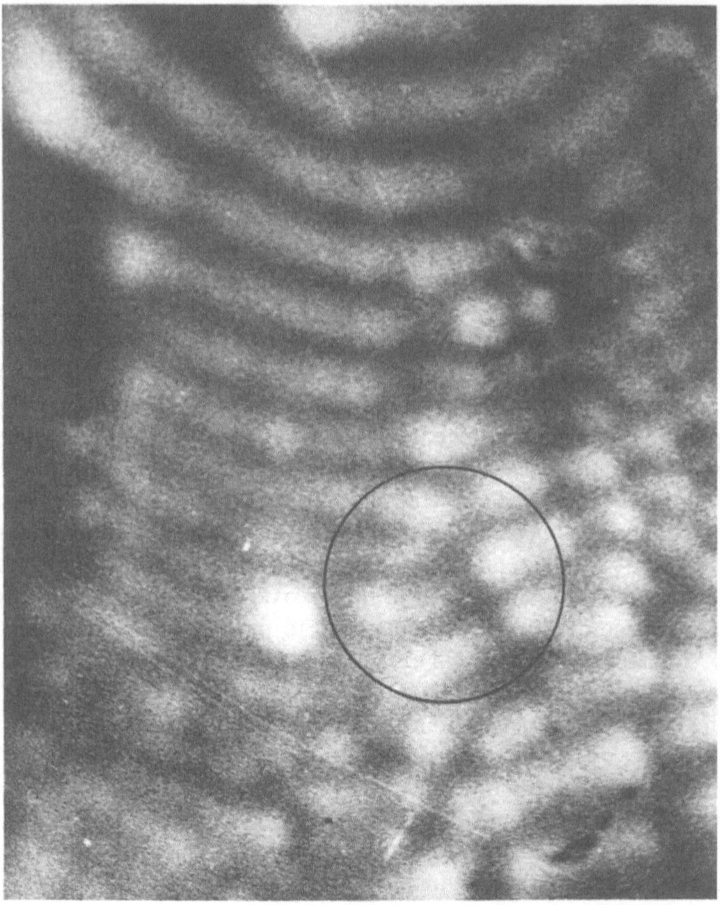

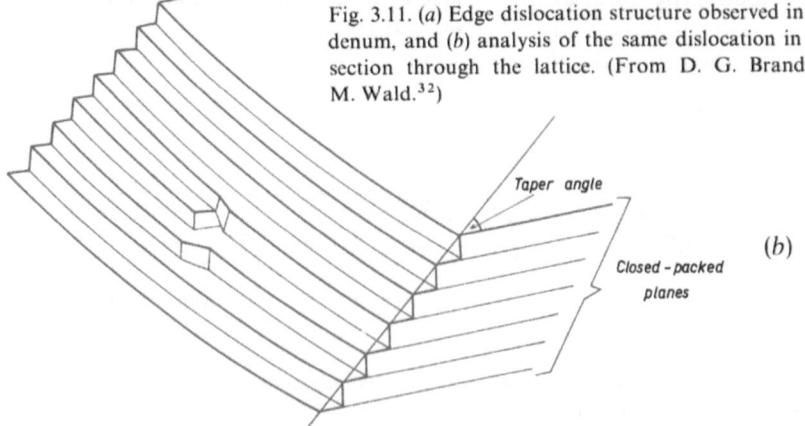

(a)

(b)

Taper angle

Closed-packed planes

Fig. 3.11. (a) Edge dislocation structure observed in molybdenum, and (b) analysis of the same dislocation in a taper section through the lattice. (From D. G. Brandon and M. Wald.[32])

While point defects are best observed on loosely packed planes, where they have no nearest neighbor atoms within the plane, dislocations are most easily identified when they emerge at the edges of the close-packed planes. In this region the field-ion image is essentially a taper section through the lattice, and the dislocation structure is magnified by the taper section and very clearly revealed[32] (Fig. 3.11). When the Burgers vector of the dislocation lies parallel to the edges of the close-packed planes, no displacement is observed in the image, but, when a component of the Burger's vector lies out of the plane of the surface, the shallow surface step produced may give rise to distortion in the image. An example in which two partial dislocations, separated by a stacking fault, produce distortion but no net displacement of the lattice planes is shown in Fig. 3.12.[33]

Remarkably little preferential evaporation occurs at grain boundaries, and this has been associated with the good fit that obtains across the boundaries usually observed.[34] Most of these boundaries are close to high-density coincidence orientations, and such preferential evaporation as does

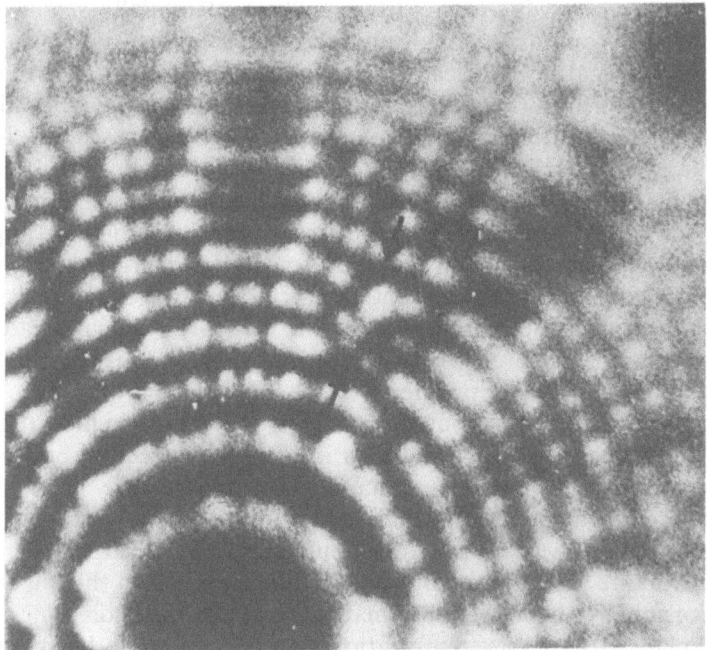

Fig. 3.12. Partial dislocations separated by a stacking fault in tungsten. (From H. Ryan and J. C. Suiter.[33])

occur is confined to misfit regions that arise from the deviation from co-incidence and from steps in the boundary. Any removal of loosely bound atoms at the boundary tends to result in group evaporation, but, even so, the width of the zone affected is generally no more than three atom diameters.

3.7   Structure of Field-Evaporated Surfaces—Alloys

In an alloy the binding energy of an atom depends not only on its position in the lattice but also on its atomic species and the atomic species of its neighbors. The ideal order of evaporation is thus no longer maintained. For very dilute alloys and for impurities in otherwise pure metals, atoms of the second component lead to local deviations from the smooth section (Fig. 3.5), which can still be used as a basis for interpretation.

If the alloy is sufficiently dilute, the discussion in Section 3.3 will apply and we need only consider atoms of the second component and their nearest neighbors. It is useful to distinguish between metallic and nonmetallic impurities. For nonmetallic impurities the major effect is not due to any difference in binding energy but to the very much higher ionization potentials of these impurities, which tend either to be preferentially retained in the lattice or to be evaporated as molecular ions. As previously noted, a tendency to surface segregation can lead to preferential evaporation of the matrix over underlying impurities. For metallic impurities and dilute alloys the difference in ionization potential of the two components plays a part in controlling the order of evaporation but is not necessarily the determining factor (compare the ionization potentials and sublimation energies listed in Table 3 for various solute elements in iron); unlike nonmetallic impurities, preferential evaporation of alloys is often expected and molecular ions should not be formed. However, preferential retention of the second component and a reduction in the binding energy of the nearest neighbor atoms can lead to surface structures which are identical with those formed in the presence of nonmetallic impurities.

At higher concentrations interactions between neighboring atoms cannot be ignored. The perturbations in the surface section become more widespread, and eventually they obscure the lattice structure.[35] In an ordered alloy preferential evaporation can lead to multiple lattice steps, but, because the second component is now distributed on a superlattice, the regular surface structure is preserved and can be interpreted from this superlattice.

Occasionally the higher sublimation energy of one component is very nearly balanced by a lower ionization potential of the other component, and a regular crystallographic surface may then be obtained after field evaporation. This seems to be the case for a platinum-10% iridium alloy.[36]

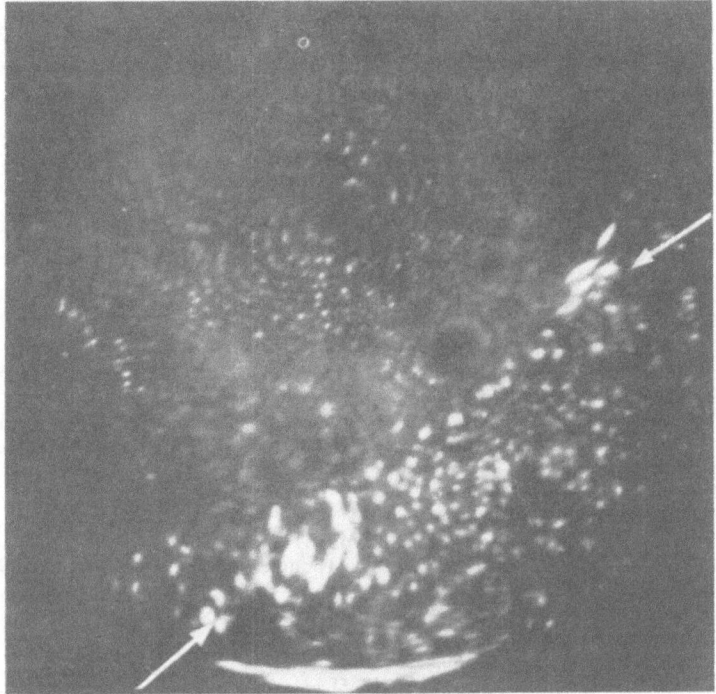

Fig. 3.13. Grain boundary in a tungsten and 5% rhenium alloy showing the contrast associated with impurity segregation. (From B. Ralph and D. G. Brandon.[35])

Segregation of impurities to lattice defects may result in very high local concentrations of impurity which partially obscure the structure of the defect. An example of such segregation at a grain boundary is shown in Fig. 3.13.

3.8 Conclusions

1. Field evaporation can be understood in terms of one-dimensional potential-energy curves.

2. Quantitative measurements of the evaporation field and of the field and temperature sensitivity of the evaporation rate are in general agreement with a simple theory based on a polarization–modified image-force law.

3. The theory is adequate for describing the evaporation of pure metals and of single isolated atoms of a second component.

4. The field-evaporated surface can be approximated by a smooth section taken through the crystal lattice.

5. The form of this section is determined by the orientation dependence of the field-evaporation process and by the tip geometry.

6. Deviations from this section result from preferential evaporation or retention of a second component or of the neighboring atoms. Deviations also arise from preferential evaporation near lattice defects.

References

1. E. W. Müller, *Phys. Rev.* **102**: 618 (1956).
2. E. W. Müller, *Advan. Electron. Electron Phys.* **13**: 83 (1960).
3. E. W. Müller, in: *Structure and Properties of Thin Films*, C. A. Neugebauer, J. D. Newkirk, and D. A. Vermilyea, eds., Wiley (New York), 1959, p. 476.
4. S. S. Brenner, in: *High-Temperature and High-Resolution Metallography*, G. S. Ansell and H. I. Aaronson, eds., Gordon & Breach (New York), 1967 [Vol. 38 of the *Met. Soc. Conf.*, (Chicago, Feb. 1965)].
5. H. Utsugi and R. Gomer, *J. Chem. Phys.* **37**: 1706 (1962).
6. H. Utsugi and R. Gomer, *J. Chem. Phys.* **37**: 1720 (1962).
7. L. W. Swanson and R. Gomer, *J. Chem. Phys.* **39**: 2813 (1963).
8. L. W. Swanson, R. W. Strayer, and F. J. Charbonnier, *Surface Science* **2**: 177 (1964).
9. R. Gomer, *J. Chem. Phys.* **31**: 341 (1959).
10. R. Gomer and L. W. Swanson, *J. Chem. Phys.* **38**: 1613 (1963).
11. D. G. Brandon, *Surface Science* **3**: 1 (1965).
12. D. G. Brandon, *Phil. Mag.* **14**: 803 (1966).
13. D. G. Brandon, *Surface Science* **5**: 137 (1966).
14. S. Glasstone, K. J. Laidler, and H. Eyring, *The Theory of Rate Processes*, McGraw Hill (New York), 1941.
15. D. G. Brandon, *Brit. J. Appl. Phys.* **14**: 474 (1963).
16. G. Ehrlich, *Brit. J. Appl. Phys.* **15**: 349 (1964).
17. O. Kubaschewski and E. Ll. Evans, *Metallurgical Thermochemistry*, Pergamon Press (London), 1958.
18. O. Kubaschewski, *Landolt–Bornstein, Tables* II, 2(b) (1962), p. 2–1.
19. O. Kubaschewski, *Landolt–Bornstein, Tables* II, 4 (1962), p. 837.
20. J. F. Mulson and E. W. Müller, *J. Chem. Phys.* **38**: 2615 (1963).
21. R. Gomer, *Field Emission and Field Ionization*, Harvard University Press (Cambridge, Mass.), 1961, p. 45.
22. W. P. Dyke and W. W. Dolan, *Advan. Electron. Electron Phys.* **8**: 89 (1956).
23. M. Drechsler and P. Wolf, *Proc. 4th Intern. Conf. Electron Microscopy, Berlin, 1958*, Vol. 1, p. 835.
24. R. Gomer, *Field Emission and Field Ionization*, Harvard University Press (Cambridge, Mass.), 1961, p. 47.
25. E. W. Müller and R. D. Young, *J. Appl. Phys.* **32**: 2425 (1961).
26. A. G. J. van Oostrom, Doctoral thesis, University of Amsterdam, 1965, p. 67.
27. R. Smoluchowski, *Phys. Rev.* **60**: 661 (1941).
28. J. Friedel, *Advan. Phys.* **3**: 446 (1954).
29. A. J. W. Moore, *J. Phys. Chem. Solids* **23**: 907 (1962).
30. E. W. Müller, *Surface Science* **2**: 484 (1964).
31. M. K. Sinha and E. W. Müller, *J. Appl. Phys.* **35**: 1256 (1964).
32. D. G. Brandon and M. Wald, *Phil. Mag.* **6**: 1035 (1961).
33. H. Ryan and J. C. Suiter, *J. Less-Common Metals*, (1965).
34. D. G. Brandon, S. Ranganathan, B. Ralph, and M. Wald, *Acta Met.* **12**: 813 (1964).
35. B. Ralph and D. G. Brandon, *Phil. Mag.* **8**: 919 (1963).
36. E. W. Müller, *J. Phys. Soc. Japan* **18** (II): 1 (1963).

Chapter 4

GAS IMPACT, FIELD ETCHING, AND FIELD DEFORMATION

D. G. Brandon†

4.1 Gas Arrival Rate

The rate of arrival of gas molecules at a specimen tip is enhanced by polarization effects at moderate field strengths, but at high field strengths field ionization of the gas in space drastically reduces the net arrival rate. The gas kinetic arrival rate per unit area, n_0, is given by

$$n_0 = p(2\pi MkT)^{-\frac{1}{2}} \tag{4.1}$$

where p is the gas pressure; M, the molecular weight of the gas; and T, the gas temperature in °K. If p is measured in μtorr equation (4.1) becomes

$$n_0 = 3.54 \times 10^{16} p(MT)^{-\frac{1}{2}} \tag{4.2}$$

When an electric field F_0 is applied to a spherical emitter, the gas supply is enhanced by a factor ξ given by[1]

$$\xi = \frac{\mu F_0}{kT} + \left(\frac{1/2\pi\alpha F_0^2}{kT}\right)^{\frac{1}{2}} \text{erf}\left[\left(\frac{1/2\alpha F_0^2}{kT}\right)^{\frac{1}{2}}\right] \tag{4.3}$$

where μ is the permanent dipole moment of the gas and α its polarizability. For $x > 2$, erf $x \cong 1$, and the last term in equation (4.3) then reduces to

$$\left(\frac{\frac{1}{2}\pi\alpha F_0^2}{kT}\right)^{\frac{1}{2}}$$

The probability P that a gas molecule will be ionized on its way to the tip is given by[2]

$$P = 1 - \exp\left(-\int_0^{t} \frac{dt}{\tau}\right) \tag{4.4}$$

---

† Senior Research Scientist, Battelle Memorial Institute, Geneva, Switzerland. Presently Professor of Metallurgy, The Technion, Israel Institute of Technology, Haifa, Israel.

where τ is the characteristic time required to ionize the molecule at a distance r from the tip center. The time τ depends on the tunneling probability $\Delta(l)$ and the characteristic orbital frequency v of the tunneling electron,[3,4]

$$\tau^{-1} = v\,\Delta(l) \cong \frac{Fev}{\frac{4}{3}(2m/\hbar^2)^{\frac{1}{2}}I^{\frac{3}{2}}} \exp\left[-\frac{4}{3}\left(\frac{2m}{\hbar^2}\right)^{\frac{1}{2}}\frac{I^{\frac{3}{2}}}{Fe}\right] \tag{4.5}$$

where F is the field strength at a distance r from the tip center, I is the ionization potential of the gas, m is the mass of the electron, e is the charge on the electron, and $\hbar$ is Planck's constant divided by 2π. If F is in volts per angstrom and I in electron volts, equation (4.5) reduces to

$$\tau^{-1} = \frac{Fv}{0.68I^{\frac{1}{2}}} \exp\left(-0.68I^{\frac{3}{2}}/F\right) \tag{4.5a}$$

If R is the tip radius and $r > R$, then F near the tip is given approximately by[4]

$$F = F_0\left(\frac{R}{r}\right)^2 \tag{4.6}$$

so that

$$\frac{dr}{dF} = -\frac{r}{2F} \tag{4.7}$$

If the transverse component of velocity and the thermal energy are neglected, the kinetic energy of a molecule arriving from infinity is just equal to the drop in potential energy arising from polarization in the electric field

$$\frac{1}{2}M\left(\frac{dr}{dt}\right)^2 = \mu F + \frac{1}{2}\alpha F^2 \tag{4.8}$$

or

$$\frac{dr}{dt} = -\left[\frac{2(\mu F + 1/2\alpha F^2)}{M}\right]^{\frac{1}{2}} \tag{4.8a}$$

the negative root being chosen because the molecule is moving *toward* the tip.

By combining equations (4.6), (4.7), and (4.8)

$$\frac{dt}{dF} = \frac{R}{2F}\sqrt{\frac{F_0 M}{2F(\mu F + 1/2\alpha F^2)}} \tag{4.9}$$

By setting $u = \int_0^t \frac{dt}{\tau}$ and substituting for τ and dt:

$$u = \int_0^{F_0} \frac{Fv}{0.68I^{\frac{1}{4}}} \exp \frac{-0.68I^{\frac{3}{4}}}{F} \frac{R}{2F} \sqrt{\frac{F_0M}{2F(\mu F + 1/2\alpha F^2)}} \, dF \qquad (4.10)$$

If we write $x = \frac{4}{3}(2m/\hbar^2)I^{\frac{3}{4}}/Fe = a/F$ (so that we transform the expression to a dimensionless variable) and assume $\mu F \ll 1/2\alpha F^2$ (which is generally true at high field strengths),

$$u = \frac{Rv}{2a} \sqrt{\frac{F_0M}{\alpha a}} \int_{x_0}^{\infty} -x^{-\frac{1}{2}}e^{-x} \, dx \qquad (4.11)$$

We are interested in the range of F_0 from 1 to 7 V/Å corresponding to a range of x_0 (taking I as 15 eV) of 40 to 5.

Repeated integration by parts gives

$$\int_{x_0}^{\infty} -x^{-\frac{1}{2}}e^{-x} \, dx = x_0^{-\frac{1}{2}}e^{-x_0}\left(1 - \frac{1}{2}\frac{1}{x_0} + \frac{3}{4}\frac{1}{x_0^2} - \frac{15}{8}\frac{1}{x_0^3} + \right) \cong x_0^{-\frac{1}{2}}e^{-x_0}$$

in the range of interest.

The total supply of gas per unit area n is obtained by combining equations (4.1), (4.3), and (4.4) thus

$$n = n_0\xi(1 - P) \qquad (4.12)$$

i.e., the number of atoms striking the tip per unit area is given by the product of the enhancement factor, the gas kinetic arrival rate, and the probability that a gas atom will escape ionization.

We can rewrite equation (4.12), by using equation (4.11), as

$$\ln\frac{n}{n_0} = \ln\left\{\mu\frac{F_0}{kT} + \left[\frac{\pi(1/2\alpha F_0^2)}{kT}\right]^{\frac{1}{2}}\right\} - \frac{Rv}{2a}\sqrt{\frac{F_0M}{\alpha a}}x_0^{-\frac{1}{2}}e^{-x_0} \qquad (4.12a)$$

which is of the form[5]

$$\log\frac{n}{n_0} = \log(AF_0) - BF_0\exp\left(\frac{-a}{F_0}\right) \qquad (4.13)$$

where the first term on the right gives the polarization enhancement factor and the second term the losses through field ionization.

Some calculated curves for the rate of arrival of contaminant gas species are given in Fig. 4.1a and may be compared with the observed contamination rate for a tungsten specimen as a function of applied field strength,* shown in Fig. 4.1b. The general form of equation (4.13) is obviously correct, although the quantitative agreement is poor, probably because only a proportion of the impinging gas sticks to the surface.

* Southon, private communication.

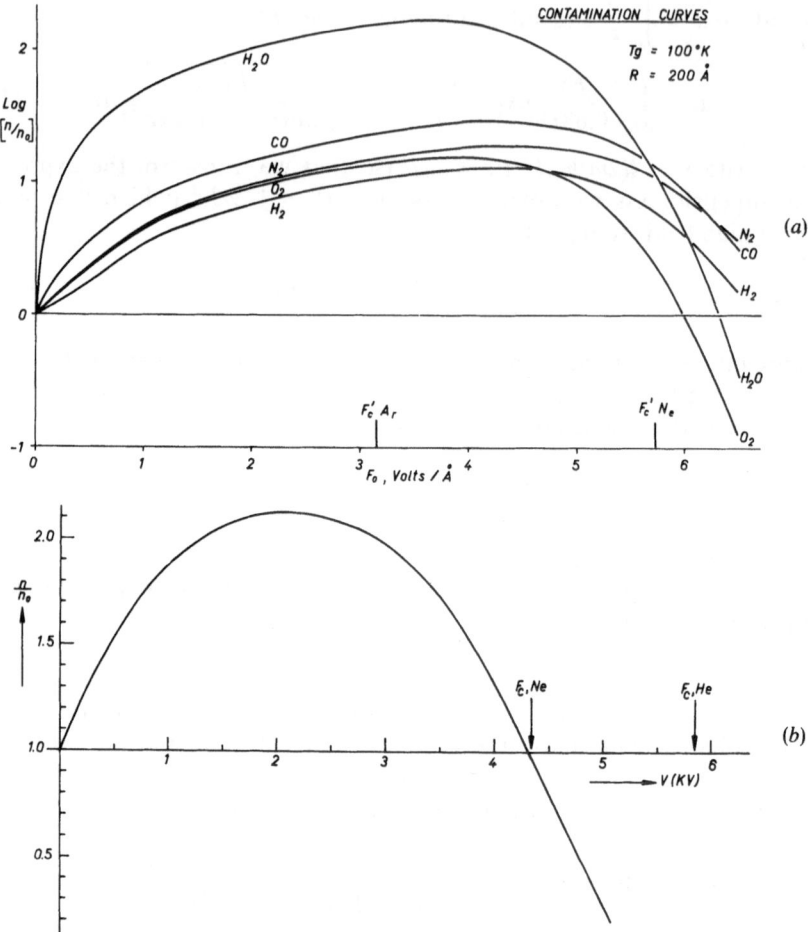

Fig. 4.1. (a) Calculated gas arrival rate for common contaminant species; (b) experimentally observed contamination rate as a function of field strength. (Southon, private communication.)

4.2 Energy Transfer in Gas Impact

The classical definition for the accommodation coefficient a of a gas atom colliding with an atom at a free surface is given by

$$a = \frac{E_0 - E_1}{E_0 - E_s} \tag{4.14}$$

where E_0 is the initial kinetic energy of the impinging gas atom, E_s is the initial kinetic energy of the struck atom, and E_1 is the kinetic energy of the gas atom after the collision.

For hard-sphere collisions the maximum energy which can be transferred, E_{max}, is given by

$$E_{max} = (E_0 - E_1)_{max} = \frac{4m^*}{(m^* + 1)^2}\left[(E_0 - E_s) + \frac{(m^* - 1)}{\sqrt{m^*}}\sqrt{E_0 E_s}\right] \quad (4.15)$$

where m^* is the ratio of the atomic mass of the struck atom to that of the impinging gas atom.

When $E_s = 0$ equation (4.15) reduces to the well-known relation

$$E_{max}^0 = \frac{4m^*}{(m^* + 1)^2}E_0 \quad (4.16)$$

The probability P that an energy greater than Q will be transferred to the struck atom in a hard-sphere collision is given approximately by

$$P = 1 - \frac{Q}{E_{max}} \quad (4.17)$$

$$(Q = E_0 - E_1 \leqq E_{max})$$

if the struck atoms are arranged in a regular array. For $E_s = 0$, the mean value of a, $\bar{a}_0$, is then given by

$$\bar{a}_0 = \frac{2m^*}{(m^* + 1)^2} \quad (4.18)$$

At finite temperatures $(E_s \neq 0)$ a fraction of the impinging atoms will be able to transfer energies greater than E_{max}^0 to the lattice. Calculations of the correct distribution function for the accommodation coefficient[6,7] lead to the conclusion that a good approximation to this high-energy tail in the distribution of energies transferred is given by a Maxwell–Boltzman function with an effective temperature T^*

$$kT^* = \bar{a}_0 kT_g + (1 - \bar{a}_0)kT \quad (4.19)$$

where $\bar{a}_0$ is given by equation (4.18) and T_g is the effective gas temperature at the tip.

In the following section we shall use equations (4.17) and (4.19) to estimate the effect of inert-gas impact on field evaporation.

4.3 Impact-Promoted Field Evaporation

In the presence of an inert gas, such as the image gas in the field-ion microscope, field evaporation occurs at a slightly reduced field strength,[8] and the field strength and temperature dependence of the evaporation rate are also reduced.[9] These effects can be explained by the kinetic energy transfer from the inert gas arriving at the tip to the surface atoms. The effective bombardment frequency for the surface atoms, v^*, is given by

$$v^* = nv^{\frac{2}{3}} \qquad (4.20)$$

where n is defined by equations (4.1) and (4.12) and v is the atomic volume of the metal.

To a first approximation we can assume the probability of activation in a single collision event to be given by equation (4.17) with

$$E_{\max} \cong a_0 \frac{1}{2} \alpha F_0^2 = \frac{2m^* \alpha F_0^2}{(m^* + 1)^2} \qquad (4.21)$$

The evaporation rate under these conditions is given by

$$k_e = v^* \left[1 - \frac{(m^* + 1)^2 Q}{2m^* \alpha F_0^2} \right] \qquad (4.22)$$

Assuming that field ionization losses in the bombarding gas are negligible and that there is no permanent dipole moment, we have from equations (4.1), (4.12), and (4.20)

$$v^* = \frac{4.43 v^{\frac{2}{3}}}{\sqrt{MT_g}} p \sqrt{\frac{\alpha F_0^2}{kT_g}} \qquad (4.23)$$

where p is the gas pressure in μtorr, T_g is now the gas temperature far from the tip, and v is in cubic angstroms.

Combining equations (4.22) and (4.23) and taking $Q \sim Q_0 - (n^3 e^3 F_0)^{\frac{1}{2}}$ [equation (3.3)], we can differentiate k_e with respect to F_0 at constant pressure and temperature. Remembering

$$Q \ll Q_0 \qquad \text{and} \qquad \frac{2m^*}{(m^* + 1)^2} \alpha F_0^2 = E_{\max}^0$$

we then find

$$\left(\frac{\partial \ln k_e}{\partial \ln F_0} \right)_{T_g, p} = \frac{\frac{1}{2}(Q_0 + 4Q)}{(E_{\max}^0 - Q)} + 1 \qquad (4.24)$$

Comparison with thermally activated field evaporation [equation (3.13)] shows that gas-impact-promoted field evaporation replaces kT by $E_{\max}^0 - Q$ and increases Q_0 by $4Q$, the latter due to the field dependence of $E_{\max}^0$. The factor 1 on the right of equation (4.24) arises from the linear field dependence of v^* [equation (4.12)]. The *reduction* in the field sensitivity of the evaporation

TABLE 1

Field-Evaporation Parameters
0.2 μtorr Ne, 63°K

	Ψ63°K Observed	$\left(\dfrac{kT}{E^{\circ}_{\max}}\right)_{T=63°K}$	$\left(\dfrac{\partial \ln k_e}{\partial \ln F}\right)_{T_g p}$	$\left(\dfrac{\partial \ln p}{\partial \ln F}\right)_{k_e T_g}$
W	0.63	0.03	236	180
Mo	....	0.03	304	175
Pt	0.71	0.14	173	150

rate arising from the impact of an inert bombarding gas can be expressed by a factor ψ

$$\psi \cong \frac{kT}{E^0_{\max}} \tag{4.25}$$

when $Q \ll E^0_{\max}$.

For neon on tungsten $m^* = 9.1$ and $\alpha = 0.392$ Å^3. At the evaporation field for tungsten, 6.8 V/Å, $1/2\alpha F_0^2 = 0.60$ eV, so that, if T is 77°K, $\psi \cong 0.03$. Observed values of ψ (Table 1) are much greater than this.

If we use equation (4.19) to describe the distribution of energies transferred to the tip by gas impact and assume that kT depends on F_0 in the same way as $E^0_{\max}$, equation (4.24) becomes

$$\left(\frac{\partial \ln k_e}{\partial \ln F_0}\right)_{T_g p} = \frac{\frac{1}{2}Q_0}{kT^*} + 2 \ln \frac{v^*}{k_e} + 1 \tag{4.26}$$

If we take $kT_g \cong \frac{1}{2}E_0$ [equations (4.14) and (4.19)] and use the figures given for neon above, this yields $\psi \cong 0.12$, in better agreement with the experimental values.

Since $v^* \propto p$ [equation (4.23)], by keeping k_e/p constant the observed evaporation field should be kept constant. This has indeed been observed over a wide range of gas pressures (Fig. 4.2).[8] Also, since k_e is linearly dependent on p, the pressure sensitivity of the evaporation field at constant k_e is inversely proportional to the field sensitivity of the evaporation rate at constant p

$$\left(\frac{\partial \ln p}{\partial \ln F_0}\right)_{T_g k_e} = -\left(\frac{\partial \ln k_e}{\partial \ln F_0}\right)_{T_g p} \tag{4.27}$$

Because of changes in tip size during the experiment it is difficult to measure $(\partial \ln p/\partial \ln F_0)_{k_e T_g}$, but approximate values are in reasonable agreement with equation (4.27) (Table 1).

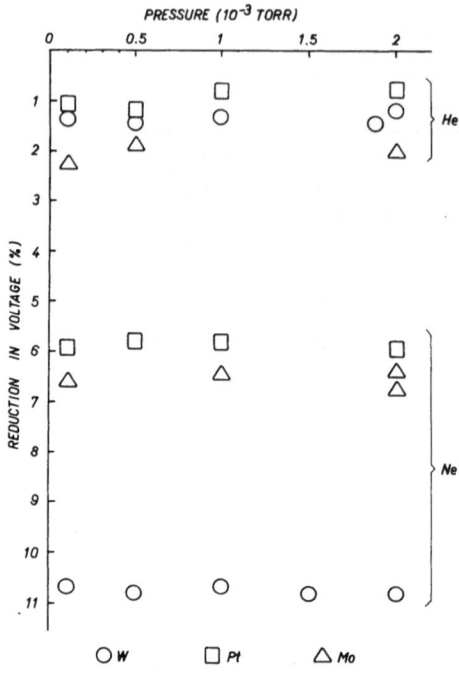

Fig. 4.2. Reduction in evaporation field under gas impact at constant k_e/p. (From O. Nishikawa and E. W. Müller.[8])

4.4 Field Etching

As $k_e \rightarrow v^*$, the rate of gas-impact-promoted evaporation will become limited by the arrival rate of the gas, and the field sensitivity of the evaporation rate should then fall drastically. This is analogous to the stable-current region of the field ionization curve. In gas-impact-promoted field evaporation this second regime has not been observed experimentally since thermally activated field evaporation takes over unless E_{max}^0 is a large fraction of Q_0.

In the presence of a chemically active gas, field evaporation is often observed at a very low field strength (a phenomenon known as *field etching*), and this much-reduced field sensitivity of the evaporation rate is observed. The field sensitivity must then be attributed to a supply function-limited evaporation process, $k_e \cong v^*$, where v^* is given by equations (4.1), (4.12), and (4.20).

Figure 4.3 shows schematically the expected dependence of the evaporation rate on the applied field strength for (a) gas-impact-promoted field evaporation and (b) field etching. In (b) it has been assumed that at a low enough field strength the evaporation will depend on energy transfer from the incoming gas (stage I). At intermediate field strengths the evaporation rate depends on the gas supply rate n (stage II). At still higher field strengths

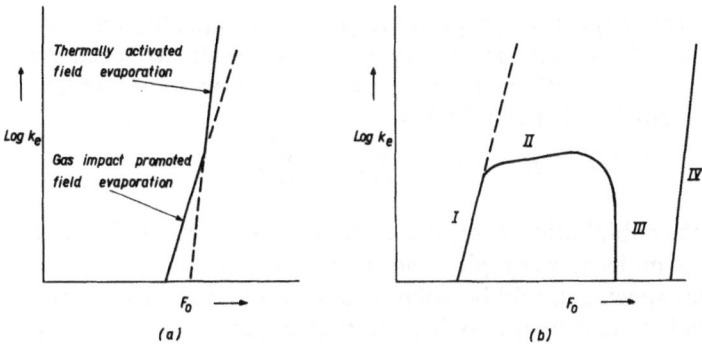

Fig. 4.3. Expected dependence of evaporation rate on applied field strength for (a) gas-impact-promoted field evaporation and (b) field etching (see text).

the gas is ionized in space before reaching the tip, so that the observed evaporation rate falls rapidly (stage III, *cf.* Fig. 4.1). Finally, thermally activated field evaporation takes over (stage IV). In gas-impact-promoted field evaporation (Fig. 4.3a) the first and final stages of Fig. 4.3b generally blend into one another. In field etching the first stage has not been observed in practice and may be absent because of surface contamination effects. Stage III will also be absent if thermally activated evaporation intervenes before field ionization in space seriously diminishes the incoming gas supply.

To explain field etching it is necessary to assume that Q_0 has been significantly reduced by the presence of the active gas. Only if this is true will the condition $k_e \cong v^*$ be approached at field strengths significantly below the normal evaporation field. It is thus natural to attempt to explain field etching in terms of the evaporation of a molecular complex with a much reduced Q_0 compared with that of the matrix metal (*cf.* Section 3.3).

We therefore wish to compare the following possibilities:[10]

$$M_s \longrightarrow M^{n+} + ne_m - Q_0^M \tag{a}$$

$$M_s + A_{ads} \longrightarrow M_s + A_{ads}^{n+} + ne_m - Q_0^A \tag{b}$$

and

$$M_s + A_{ads} \longrightarrow MA^{n+} + ne_m - Q_0^C \tag{c}$$

where the subscript s refers to the solid metal, the superscripts M, A, and C refer to the metal, the active gas, and the molecular complex respectively, and e_m refers to an electron returned to the metal. In addition we may wish to consider the circumstance in which the incoming gas is not adsorbed in the ground state but is evaporated as a molecular complex before the heat of adsorption has been dissipated

$$M_s + A \longrightarrow MA^{n+} + ne_m - Q_0^{*C} \tag{d}$$

The energy schemes appropriate to possibilities (a), (b), (c), and (d) are shown in Fig. 4.4. Possibilities (a), (b), and (c) have already been discussed in Chapter 3. The condition under which (c) would occur at a reduced field strength compared with that of (a) is:

$$\Lambda_C - \Lambda_M < \sum_n I_M - \sum_n I_C \tag{4.28}$$

where Λ is the sublimation energy, I is the ionization potential, and C and M refer to the molecular complex and the metal as before. This implies that the molecular species should be volatile and should also have a low ionization potential. It is difficult to believe that equation (4.28) is fulfilled in practice. The fourth possibility (d) implies that the heat of adsorption H_a should be subtracted from Λ_C. We thus have

$$Q_0^M = \Lambda_M + \sum_n I_M - n\phi \tag{i}$$

$$Q_0^A = H_a + E_d + \sum_n I_A - n\phi \tag{ii}$$

$$\text{(4.29)}$$

$$Q_0^C = \Lambda_C + \sum_n I_C - n\phi \tag{iii}$$

$$Q_0^{*C} = \Lambda_C - H_a + \sum_n I_C - n\phi \tag{iv}$$

From Fig. 4.4 and equations (4.29) we see that the fourth possibility (d) is the one that best accounts for the field etching process. Field etching is

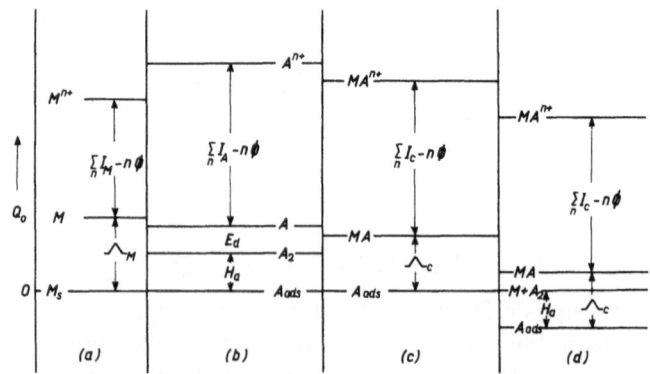

Fig. 4.4. Energy schemes for field evaporation in the presence of an active gas: (a) normal field evaporation of the metal, (b) field evaporation of the adsorbed gas, (c) field evaporation of the stable molecular complex, and (d) field evaporation of a molecular complex in an excited state.

expected whenever the heat of adsorption is large enough to compensate for the relatively high ionization potential one might expect for a molecular ion.

So far field etching has only been observed in the presence of the common contaminant gases, H_2, CO, and N_2.[11] Because of the volatility of the halides it would be of great interest to observe field etching behavior in the presence of the halogens, especially fluorine, which has both a high ionization potential (17.3 eV) and a low boiling point (85°K).

Since field etching is relatively insensitive to the applied field strength and occurs at a low field strength, this process can be used as an alternative to thermally activated field evaporation for those metals that deform at the field strengths required for the latter process.[12] The technique has been very successfully applied to the preparation of niobium, iron, and nickel specimens with hydrogen used as the active gas.[13]

The effect of an active gas on the field evaporation of nonmetallic impurities is also important. Preferential field etching of the impurity is possible if a strong bond is formed with the active gas. Since suitable data on ionization potentials and heats of adsorption are often available for molecular compounds of the nonmetallic elements, the effect of the active gas on the evaporation process can sometimes be calculated numerically. As an example, consider the effect of hydrogen on the field evaporation of nitrogen adsorbed on iron. We wish to compare

$$N_{sol} \rightarrow N^+ + e_m - Q_0^N \tag{e}$$

with

$$N_{sol} + H_2 \rightarrow NH_2^+ + e_m - Q_0^{*NH_2} \tag{f}$$

From the literature

$$\tfrac{1}{2}N_2 \rightarrow N_{sol} - 1.23 \text{ eV} \tag{g}$$

$$\tfrac{1}{2}N_2 \rightarrow N - 3.72 \text{ eV} \tag{h}$$

$$N \rightarrow N^+ - 14.54 \text{ eV} \tag{i}$$

$$\tfrac{1}{2}N_2 + \tfrac{3}{2}H_2 \rightarrow NH_3 + 0.48 \text{ eV} \tag{k}$$

$$NH_3 \rightarrow NH_2^+ + H - 12.0 \text{ eV} \tag{l}$$

$$\tfrac{1}{2}H_2 \rightarrow H - 2.26 \text{ eV} \tag{m}$$

TABLE 2

Field Etching of Nonmetallic Impurities in Iron
$$\phi = 4.17 \text{ eV}$$

Impurity	Active gas	Q'_n, eV	Ionic species
Fe	...	7.19	Fe^{++}
C	...	11.9	C^{++}
C	H_2	9.9	CH_2^+
C	H_2	7.23	CH_4^+
C	O_2	4.62	CO_2^+
N	...	13.4	N^+
N	H_2	4.47	NH_2^+
N	O_2	6.56	NO_2^+
O	...	13.6	O^+
O	H_2	7.19	H_2O^+
O	CO	8.23	CO_2^+

Hence by suitable manipulation

$$Q_0^N - Q_0^{*NH_2} = 8.4 \text{ eV}$$

and by inserting $\phi = 4.17 \text{ eV}$

$$Q_0^{*NH_2} = 4.47 \text{ eV}$$

which corresponds to an evaporation field of about 1.4 V/Å, well below the evaporation field strength for iron ($\cong$ 3.6 V/Å).

Similar calculations are possible for a large number of active gas-impurity combinations, and some results for iron are summarized in Table 2. Clearly field etching could be a very useful technique for the selective removal of alloy constituents from a metal surface.

4.5 Field Deformation

The applied electric field generates a hydrostatic tension σ_N normal to the surface of the specimen given by[14]

$$\sigma_N = \frac{F_0^2}{8\pi} \tag{4.30}$$

or

$$\sigma_N = 44.3 F_0^2 \text{ kg-mm}^2$$

where F_0 is in volts per angstrom.

The evaporation fields of most metals of interest lie between 3.5 and 6.5 V/Å (Chapter 3), so that during field evaporation the stress normal to the

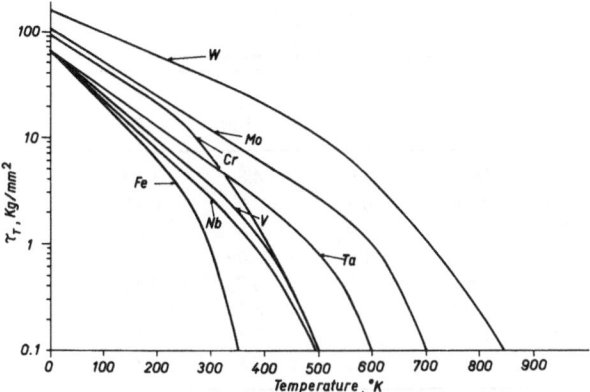

Fig. 4.5. Critical shear stress for the common body-centered cubic metals as a function of temperature.

surface is of the order of 500 to 2000 kg-mm². Most body-centered cubic metals deform plastically at shear stresses well below 100 kg-mm², even at low temperatures[15] (Fig. 4.5), so that it is at first sight surprising that any metal is capable of being field evaporated without simultaneous plastic deformation and eventual fracture. The explanation lies in the distribution of the stresses generated and the small size of the specimen. A spherical specimen of an isotropic metal will experience a uniform dilatation as a result of the applied electrostatic field. The *shear* stresses are then zero.

For a hemispherical tip superimposed on a cylindrical shank the tensile stress in the shank is approximately equal to σ_N, so that the maximum shear stress in the *shank*, τ, is given by

$$\tau \cong \frac{\sigma_N}{2} \tag{4.31}$$

These shear stresses are very large, and, unless the taper angle of the shank is large, appreciable plastic deformation may occur in the shank. Thus it is observed that, after field evaporating a platinum specimen at 20°K, subsequent annealing of the specimen at 300°K in the absence of a field leads to appreciable deformation of the tip[16] due to stress relaxation of the previously deformed shank. Extrapolated values of the shear strength at 0°K, τ_0, can be found from Fig. 4.5 and compared with τ_e, the value of τ expected at the evaporation field. These values are given in Table 3, the shear strengths being quoted in terms of the shear modulus μ. The fact that many metals are stable during field evaporation even though $\tau_e \gg \tau_0$ strongly suggests that the small specimen dimensions and the consequent absence of suitable dislocation sources must be primarily responsible for the apparent strength of these samples. This is in accord with explanations usually advanced to explain the shear strength of fine whiskers.

TABLE 3
Shear Stress Parameters

	τ_0/μ $\times$ 100	τ_e/μ $\times$ 100	$A\dagger$
W	1.0	4.8	1.00
Mo	0.7	2.6	0.78
Ta	1.0	6.8	...
Nb	1.6	7.0	...
Fe	0.7	3.7	2.42
Pt	...	6.0	2.46
Ni	...	3.4	2.52
Cu	...	4.6	3.21

$\dagger A$ = anisotropy factor.

If the specimen is not spherical or is elastically anisotropic, there will be an appreciable component of shear stress in the end cap of the specimen as well as in the shank. The anisotropy factors for some common refractory metals are also given in Table 3. Only tungsten is completely isotropic; iron and nickel have large anisotropy factors and are therefore expected to be very liable to deform plastically during field evaporation, as is indeed observed experimentally.[2]

Because of the high normal stresses [equation (4.30)] the strain energy stored in the tip is also large. This strain energy E is given by

$$E = \frac{1}{2}v\frac{\sigma_N^2}{K}$$

$$= 1.22 \times 10^7 v\frac{F_0^4}{K} \qquad \text{eV/atom} \tag{4.32}$$

where F_0 is in volts per angstrom, v is the atomic volume in cubic angstroms, and K is the bulk modulus in dynes per square centimeter. For tungsten we have $K = 3.12 \times 10^{12}$ dynes-cm^2 and $v = 15.9$ Å^3, so that at the measured evaporation field (6.7 V/Å)$E = 0.13$ eV. This is by no means negligible in comparison with the activation energy for low-temperature diffusion processes. For example, radiation damage induced by low-temperature bombardment with energetic particles starts to anneal at temperatures well below 100°K, which corresponds to activation energies of about 0.1 eV. Binding energies between solute atoms and vacant lattice sites are also typically of the order of 0.1 eV. Hence the annealing behavior of point defects in a field-ion microscope tip will be substantially modified by the superposition of an applied electrostatic field.

The elastic dilatation of the lattice, $\varepsilon_F \cong \sigma_{N/K}$, accompanying the application of a field can be observed directly[17]; if the neon and helium field-ion

TABLE 4
Maximum Elastic Strains

	$\dfrac{\varepsilon_F}{\%}$	$\dfrac{\varepsilon_0}{\%}$
W	4.5	1.0
Mo	2.3	0.8
Ta	4.7	1.0
Fe	3.3	0.7
Pt	2.6	
Ni	2.9	
Cu	3.0	

images of a tungsten tip are superimposed with the positive of one on the negative of the other, the displacement of the image points resulting from the increase in lattice dimensions is clearly visible. The magnitude of this dilatation for several metals at the calculated evaporation field is given in Table 4, where it is compared with the value of the elastic strain at the yield stress at $0°K$, ε_0, given in Table 3. Again, it is obvious that defect interactions and defect mobility must be substantially modified in comparison with the field-free condition.

It is interesting to compare the strain energy contribution to the surface energy of the tip with the contribution from the electrostatic stress. The surface energy γ induces a hydrostatic compression σ_c given very approximately by

$$\sigma_c = \frac{2\gamma}{R} \tag{4.33}$$

(The well-known equation for the excess pressure inside a liquid droplet.)

The net normal stress on the surface is thus $\sigma_N - \sigma_C$ and the total strain energy per atom, E_T, is [cf. equation (4.32)]

$$E_T = \frac{1}{2}\frac{v}{K}(\sigma_N - \sigma_C)^2 \tag{4.34}$$

The field stress will exactly balance the stress associated with the surface energy if [equations (4.30) and (4.35)]

$$F_0^2 = \frac{16\pi\gamma}{R_C} \tag{4.35}$$

For tungsten, $F_e = 6.7$ V/Å, and, setting $\gamma = 5000$ ergs/cm^2, we find

$$R_C \cong 50 \text{ Å}$$

For copper, $\gamma = 2450 \, \text{ergs/cm}^2$ and $F_e = 3.1 \, \text{V/Å}$, hence

$$R_C \cong 100 \, \text{Å}$$

Thus for all but the smallest tips the surface energy constitutes only a minor perturbation on the effect of the electrostatic field and can be neglected.

References

1. M. J. Southon, Ph.D. thesis, Cambridge University, 1963.
2. R. Gomer, *Field Emission and Field Ionization*, Harvard University Press (Cambridge, Mass.), 1963, p. 74.
3. E. W. Müller and K. Bahadur, *Phys. Rev.* **102**: 624 (1956).
4. R. Gomer, *Field Emission and Field Ionization*, Harvard University Press (Cambridge, Mass.), 1963, p. 70.
5. D. G. Brandon, in: *High-Temperature and High-Resolution Metallography*, G. S. Ansell and H. I. Aaronson, eds. Gordon & Breach (New York), 1967 [Vol. 38 of the *Met. Soc. Conf.*, (Chicago, Feb. 1965)].
6. D. M. Gilbey, *J. Phys. Chem. Solids* **23**: 1453 (1962).
7. F. O. Goodman, *J. Phys. Chem. Solids* **26**: 85 (1965).
8. O. Nishikawa and E. W. Müller, *J. Appl. Phys.* **35**: 2806 (1964).
9. D. G. Brandon, *Phil. Mag.* **14**: 803 (1966).
10. D. G. Brandon, *Surface Science* **5**: 137 (1966).
11. J. F. Mulson and E. W. Müller, *J. Chem. Phys.* **38**: 2615 (1963).
12. E. W. Müller, *Surface Science* **2**: 484 (1964).
13. E. W. Müller, S. Nakamura, O. Nishikawa, and S. B. McLane, *J. Appl. Phys.* **36**: 2496 (1965).
14. E. W. Müller, *Acta Met.* **6**: 620 (1958).
15. H. Conrad, *Relation between Structure and Mechanical Properties of Metals*, Her Majesty's Stationery Office (London), 1963, p. 475.
16. E. W. Müller, *Advan. Electron. Electron Phys.* **13**: 83 (1960).
17. E. W. Müller, K. Rendulic, and S. Nakamura, private communication.

Chapter 5

SOME GEOMETRICAL ASPECTS OF SURFACES RELATED TO FIELD-ION MICROSCOPY

A. J. W. Moore†

5.1 The Geometry of Flat Surfaces

Most surfaces which we encounter in practice have an extremely complicated surface structure with surface irregularities, deformed layers, oxide layers, and adsorbed gases.[1,2] Fortunately on the surface of field-emission tips all the complicating outer layers have been removed, and we can therefore interpret our patterns for a clean surface on an undeformed crystal. We shall therefore treat the crystal as an assembly of hard spheres, and we shall see what sort of arrangements of spheres we should expect at the surface.

Some of these arrangements are obvious. Thus in order to preserve the fourfold symmetry, we should expect a (100) surface of a cubic crystal to show a square array of atoms. However, if we admit the existence of surfaces of other orientations than those of the simple close-packed planes, we find that edge and kink atoms are necessarily present in order that the surface may present the correct orientation to the crystal lattice. In order to consider how the structure of such surfaces is related to their orientation, we must introduce the concept of an atomically smooth surface (*hkl*).

If a crystal lattice is cut by a plane *hkl* and all atoms whose centers fall on one side of the plane are removed, then the remaining surface is said to be *atomically flat*.

If a plane sits on the outermost atoms of an atomically flat surface, then no atom can be fitted in between that plane and the rest of the crystal.

5.1.1 The Simulation of Face-Centered and Body-Centered Cubic Lattices by Stacking Balls

The arrangements in both face-centered cubic (f.c.c.) and body-centered cubic (b.c.c.) lattices can be made by piling balls onto a square array

† Senior Scientist, Commonwealth Scientific and Industrial Research Organization, Division of Tribophysics, Melbourne, Australia.

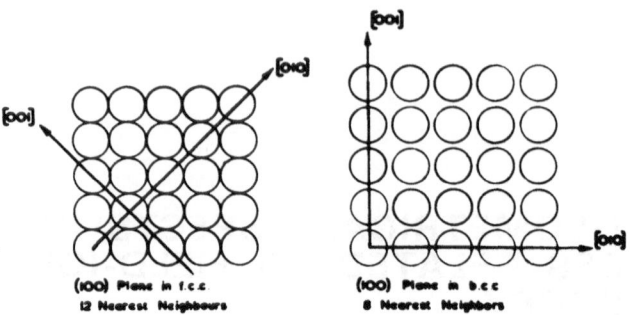

Fig. 5.1. Arrangement of balls on (100) plane in f.c.c. and b.c.c. lattice model.

representing the (100) plane, as shown in Fig. 5.1. In f.c.c. the balls in the square array are in contact and the crystal axes are at 45° to the close-packed rows. In b.c.c. the balls in the square array are $2/\sqrt{3}$ of a ball diameter apart and the crystal axes are parallel to the square array.

5.1.2 Surfaces on Face-Centered Cubic Crystals

Surfaces of orientations other than (100) may be considered as a series of steps, one interlayer distance high in the (100) plane. Thus a surface tilted at, say, $5\frac{1}{2}°$ to a (100) plane must have a step roughly every 10 atoms (tan 5° 43′ = 0.1).

For orientations along the (100)–(111) zone the step edge on an f.c.c. crystal consists of close-packed rows of atoms, and, as the orientation makes a larger angle to (100), the steps must be closer together. However, the distance between the steps can only change by integral amounts. When steps are equal distances apart, we have certain low-index orientations. When steps are as close as possible for the stability of the model, we have a tilt of 54° 44′, i.e., a (111) plane. For orientations between those where the steps are equally spaced, the distances between the steps make a sequence and they can have only two values. These are the values of the two surrounding orientations which have regular steps. The sequence is regularly repeated.

Other zones have other types of step edge. The (100)–(110) zone is at 45° to the (100)–(111) zone, and the corresponding step edge is a kinked line of atoms at 45° to the [110] axis. Again, the steps are spaced by integral distances, and, when steps are as close as gravitational stability will allow, we have a tilt of 45° to (100) and (110) planes are formed.

For other zones the step edge contains fewer kink atoms, but all orientations in the same zone contain a similar type of step edge; thus for the zone containing (100), (931), (731), (531), and (331) the step edge is a kink atom every three atoms, and intermediate orientations have two types of step displacement in a repeated sequence.

Thus in general our surface consists of a repeated sequence of kink atoms along the step edge and a repeated sequence of displacements between the steps.

5.1.3 Building a Model of Any Given (hkl) Surface

For convenience we shall specify our surface so that $h \geq k \geq l \geq 0$. The following is a brief resumé of a procedure fully described in Refs. 3 and 4.

The nature of the step edge. The step edge consists of a repeated sequence of displacements (see Fig. 5.2) in the close packed [0$\bar{1}\bar{1}$] direction (A displacements) and in the close-packed [01$\bar{1}$] direction (B displacements). The A displacements are single-atom jumps, and the B displacements may be either n jumps or $n + 1$ jumps. Examples of A and B for different zones follow.

Displacements A and B are functions of k and l. For f.c.c. they are the lowest integers which satisfy

$$\frac{A}{B} = \frac{k - l}{k + l}$$

Since $k \geq l, B \geq A$.
For (531)

$$\frac{A}{B} = \frac{2}{4} \qquad A = 1 \quad B = 2$$

For (311)

$$\frac{A}{B} = \frac{0}{2} \qquad A = 0 \quad B = 1$$

The B displacements occur in groups of n and $n + 1$ when

$$\frac{B}{A} = n + \frac{R}{A}$$

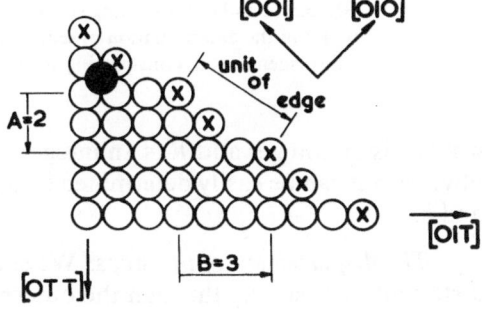

Fig. 5.2. Illustrating the nature of the displacements required to produce an arbitrary (hkl) plane with a ball-model lattice. (Parameters are discussed in Section 5.1.3.)

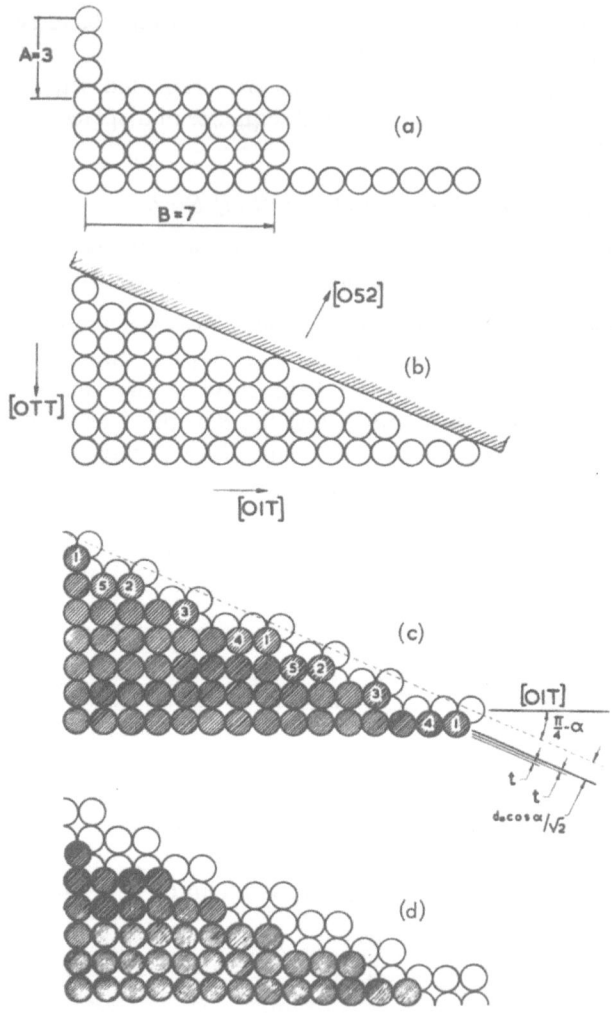

Fig. 5.3. (*a* and *b*) Illustrating step edges and use of the straight edge to aid in the determination of sequence. (*c* and *d*) The nature of the *t* displacements in constructing the plane (28 15 6).

where *n* is an integer and *R* is an integer less than *A*. The sequence is usually obvious but can be easily determined by a straight edge, as shown in Fig. 5.3*a* and *b*.

The displacement of the steps. We should now think of the step edge as a straight line passing through the outermost atoms.

When the next step is stacked as close as gravitational stability will allow, we find there is an interlayer displacement, d_0 (Fig. 5.3b).

Further displacements we shall term t displacements; they are obtained by removing the outermost atoms in the upper layer, i.e., those atoms marked 1 in Fig. 5.3c.

A repeated sequence of steps contains C, t displacements and D interlayer movements, where C and D are the smallest integers which will satisfy

$$\frac{C}{D} = B\left(\frac{h - k}{k + l}\right)$$

The distribution of the C, t displacements over the D layers is analogous to the problem of the distribution of the A and B displacements on the step edge.

Construction of model of $(hkl) = (28\ 15\ 6)$.

$$\frac{A}{B} = \frac{9}{21} \qquad A = 3 \quad B = 7$$

This is the edge already shown in Fig. 5.3b.

$$\frac{C}{D} = \frac{7 \times 13}{21} \qquad C = 13, \quad D = 3$$

i.e., there are three steps in a repeated sequence, and there are 4, 4, and 5, t displacements between them.

5, t displacements are obtained by removing the outermost atoms five times successively, i.e., atoms 1 to 5 in Fig. 5.3c and d, and then two more layers with 4, t displacements must be built up.

This gives two unit cells of the surface (28 15 6).

5.1.4 Models of Body-Centered Cubic Surface

Here the A and B displacements on the step edge are parallel to the crystal axes, as shown in Fig. 5.4.

$$\frac{A}{B} = \frac{k}{l}$$

i.e., $A \geqq B$. The B displacements occur singly. The A displacements occur in groups of n and $n + 1$ where

$$\frac{A}{B} = n + \frac{R}{B}$$

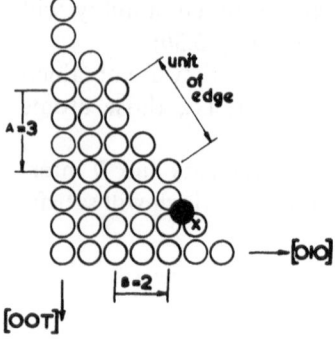

Fig. 5.4. Illustrating A and B displacements in a b.c.c. ball model. The shaded circle indicates a gravitationally unstable position sometimes required.

For the step displacements

$$\frac{C}{D} = \frac{B(h - k + l)}{2l}$$

where C is the number of t displacements and D is the number of steps, or interlayer displacements, in each repeated sequence.

Experiment with a model will show that for some orientations the interlayer displacement is obtained by having some balls in unstable gravitational positions (shaded circle in Fig. 5.4).

5.1.5 The Concentration of Different Types of Atoms on Surfaces

We can define a surface atom as one which has less than the number of nearest neighbors (n.n.) that it would have if it were in the interior of a

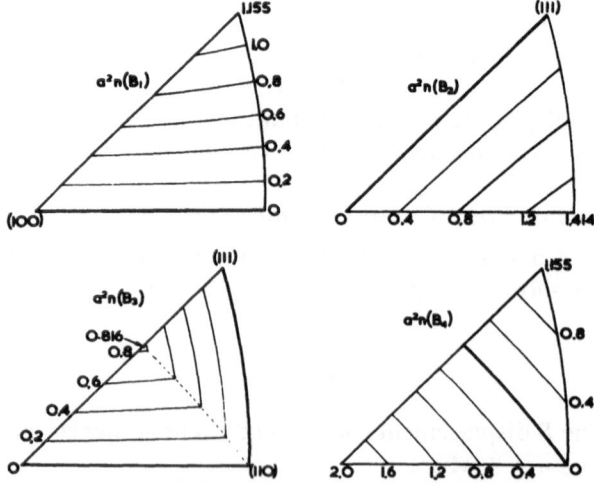

Fig. 5.5. Chart of $a^2n(Bi)$, where i is the number of broken nearest neighbor (n.n.) bonds and a is the lattice parameter for a b.c.c. lattice.

crystal, i.e., less than 12 n.n for f.c.c. and less than 8 n.n. for b.c.c. On atomically smooth f.c.c. surfaces atoms are found having 6 to 11 n.n.; on b.c.c. surfaces atoms are found with 4 to 7 n.n.

The surface concentrations of these various types of atoms vary according to the orientation, and we can construct charts showing lines of equal concentration of each type of atom.

The particular relationship between the concentration of the particular type of atom and the orientation is a simple function of (hkl), and the full procedure for determining this is given in Refs. 5, 6, and 7.

Figures 5.5 and 5.6 show the charts for b.c.c. and f.c.c. lattices. In an area a^2 of the surface (a = lattice parameter) $n(Bi)$ is the number of atoms which have i broken n.n. bonds.

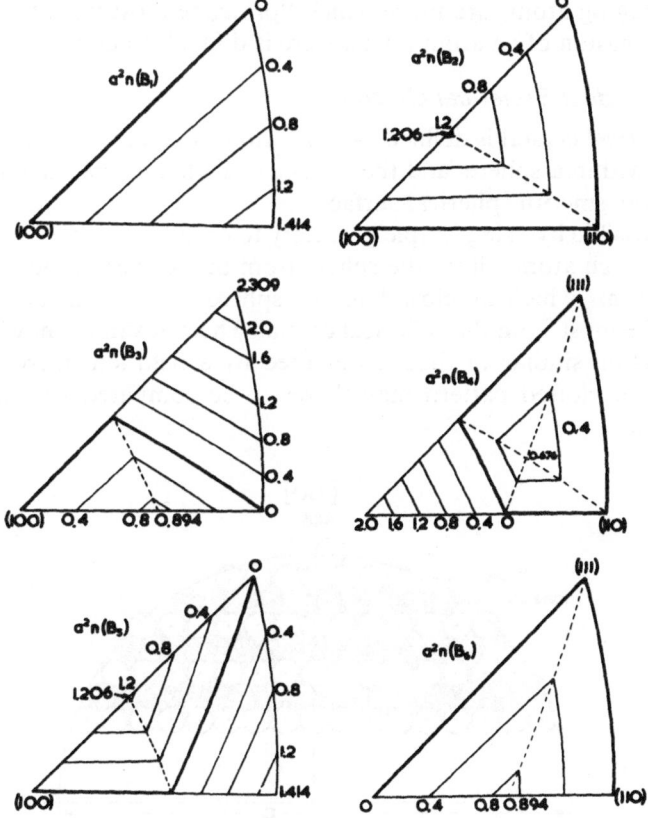

Fig. 5.6. Chart of $a^2 n(Bi)$, where i is the number of broken nearest neighbor (n.n.) bonds and a is the lattice parameter for an f.c.c. lattice.

Several interesting relationships can be observed. Kink atoms on the surface of a f.c.c. lattice have six broken bonds, and it will be noticed that they have a maximum concentration at (210). Actually one-third of all surface atoms on a (210) surface are kink atoms, and this should therefore be a very reactive surface for adsorption.

5.2 The Geometry of Spherical Surfaces

An image of a tip in a field-ion microscope is a good example of how the somewhat theoretical study of surface models could be realized in practice. The atoms near the axis of the tip are obviously arranged in circular steps which must have a varying concentration of kink atoms around them. The question immediately arises: Which atoms on the spherical tip contribute to the image? Müller gave a general answer to this problem when he said that the imaging atoms are those which "protrude most from the surface," but the protrusion of an atom on a sphere is difficult to define.[8]

5.2.1 The Ideal Spherical Crystal

A spherical crystal is defined as those atoms of an infinite crystal whose centers lie within a sphere, and the surface of such a crystal may be termed an atomically smooth spherical surface.

It is possible by using simple geometry to calculate with a computer the distance of each atom within the sphere from the surface of the sphere.[12]

The atoms which lie closest to the sphere surface will be those that protrude the most from the spherical crystal. Their positions may be plotted in a projection similar to that represented by a field-ion microscope pattern, and the plotted pattern may therefore be compared with an experimental pattern.

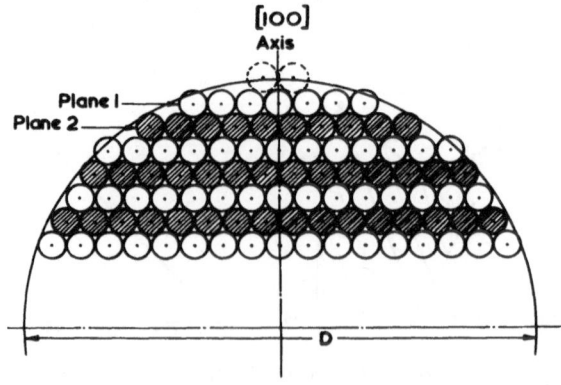

Fig. 5.7. Definition of an ideal spherical crystal (see Section 5.2.1.).

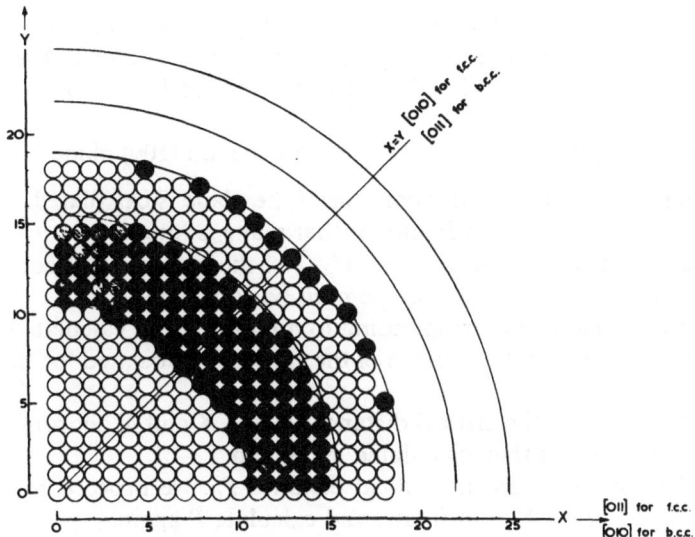

Fig. 5.8. Showing how a sphere cuts the (100) planes in an ideal spherical crystal to give approximately circular contours (see Section 5.2.1).

The center of the sphere is assumed to be on the [100] axis, which passes through lattice points on alternate (100) planes and centrally between four lattice points on the intermediate planes. See Fig. 5.7.

The surface of the sphere is made tangential to one of the latter planes, and hence all atoms in the plane are outside the sphere. The diameter of the sphere is expressed in terms of the interatomic distances in the (100) plane.

The sphere cuts the (100) planes, which give approximately circular contours, and the position of the atoms may be referred to X and Y coordinates representing the square array. See Fig. 5.8. For f.c.c. the array is close packed, and for b.c.c. the centers of the atoms are $2/\sqrt{3}$ times their diameter.

5.2.2 Outline of Computer Program

The data for the computation includes:

1. The diameter of the tip in angstroms.
2. The lattice parameter of the metal comprising the tip in angstroms.
3. The critical distance from the sphere within which it is expected that the atom will contribute to the image, the distance being expressed in interatomic distances in the (100) plane.

A mechanical plotter may be linked directly to the computer and the patterns can then be plotted rapidly and accurately. The computer takes each (100) plane in turn until sufficient of the sphere has been covered to include all orientations.

Suppose computation has just started for the nth (100) plane:

1. The radius of the circular contour of the plane is calculated.
2. The maximum X coordinate for that plane is found.
3. The maximum X coordinate is $1/\sqrt{2}$ of the maximum X coordinate (because only the basic triangle need be considered).
4. The Y coordinate corresponding to minimum X is calculated.
5. The distance P to the surface of the sphere is calculated for the atom at X, Y.
6. If P is less than the critical distance P_0, the X and Y coordinates are plotted and P is then calculated for X and $Y - 1$.
7. If P is greater than the critical distance P_0, we proceed to $X + 1$, calculate a new Y coordinate, and calculate P again.
8. The procedure is repeated until we reach the maximum X coordinate, and then the computation continues on the $n + 1$th (100) plane.

5.2.3   The Choice of Critical Distance

Suppose for an f.c.c. crystal we have a sphere radius 350 times the interatomic distance (equivalent to a Pt tip of 969-Å radius) and plot the calculated pattern for different values of the critical distance P_0 in the region of the (751) plane. This is done in Fig. 5.9. Figure 5.9a shows the effect of plotting all kink atoms and obviously provides too many image points.

The density of spots increases as P_0 increases and the plotted pattern appears to have about the same density as an experimental pattern from a tip of equivalent radius at $P_0 = 0.05$. At this value of P_0 we can construct the complete f.c.c. and b.c.c. patterns. See Figs. 5.10 and 5.11.

The similarities with experimental patterns are striking:

1. The relative importance of different planes is the same for calculated and experimental patterns.
2. The number of planes showing along any one zone is the same in both cases.

There are also differences:

1. The details on any particular plane are not the same in calculated and experimental patterns.
2. Certain bright areas do not show in the calculated pattern.

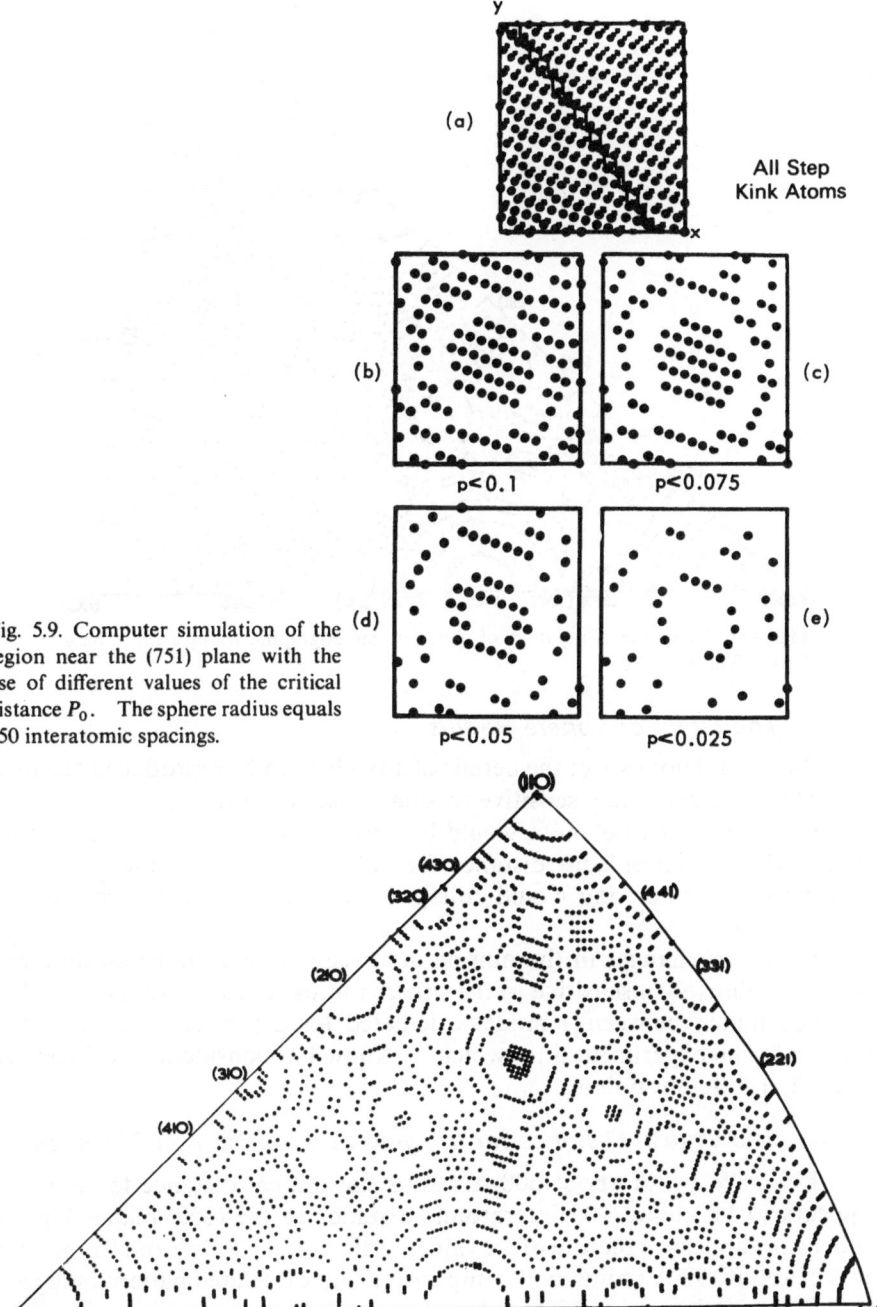

Fig. 5.9. Computer simulation of the region near the (751) plane with the use of different values of the critical distance P_0.   The sphere radius equals 350 interatomic spacings.

All Step
Kink Atoms

Fig. 5.10. Computer simulation of complete stereographic triangle for an f.c.c. lattice. The P_0 is 0.05.

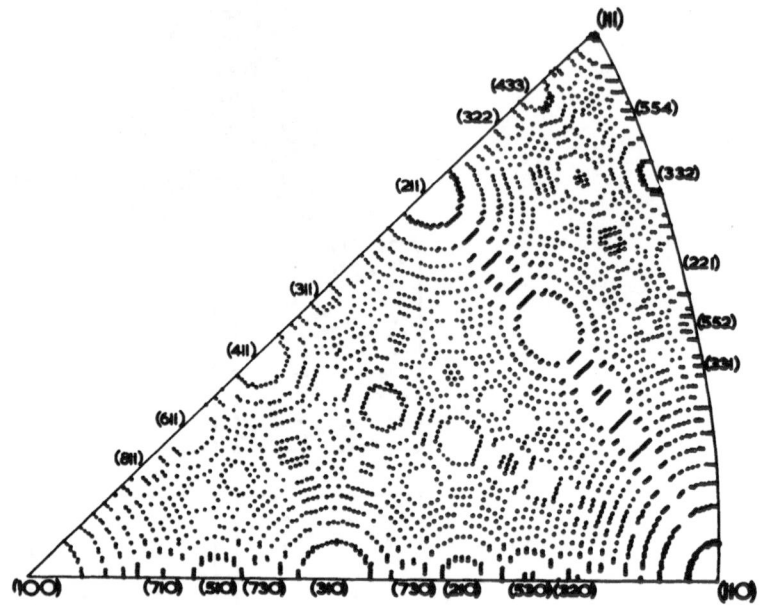

Fig. 5.11. Computer simulation of complete stereographic triangle for a b.c.c. lattice. The P_0 is 0.05.

5.2.4   The Effect of Sphere Radius

We should not expect the details of any plane to be reproduced because the details are extremely sensitive to small changes in tip radii.

For a given (*hkl*) plane it would be observed that, as the sphere radius changes by the interplanar distance for that plane, then on the pattern of that plane the inner ring would collapse and disappear to be replaced by the next ring.

However, changes in the number and relative importance of different planes is relatively insensitive to the sphere radius. Thus we can have such a startling match between a model calculated for a perfect sphere and the pattern for an experimental tip which we know has considerable differences in radii at different points.

5.2.5   The Critical Distance for Imaging as a Function of Tip Radius

For a W tip of about 600-Å radius the critical distance for imaging appears to be about 0.05 interatomic distance in the (100) plane. Tips of other sizes appear to require other values. A tip used by Ralph and Brandon[10] was of about 190-Å radius, and comparison with calculated patterns suggests that 0.1 is a more appropriate value of its P_0. Caspary and Krautz[11] used a tip of 360-Å radius, and the computed patterns suggest that $P_0 = 0.08$ is appropriate. Thus we may construct the following table:

Radius of tip, Å	Critical distance
600	0.05
360	0.08
190	0.10

Thus it would seem that, when a tip has a higher curvature, then atoms which are imbedded further below the outermost surface can contribute to the image.

5.2.6 Determination of Tip Radii

It is possible to find the tip radius between any two identifiable planes provided we can count the number of steps between them. The methods have been developed by Drechsler and Wolf[9] and Müller.[13]

Suppose we have two planes $(h_1 k_1 l_1)$ and $(h_2 k_2 l_2)$, and we can count that there are n steps of $(h_1 k_1 l_1)$ between the centers of the two planes.

We know the angle α between the planes

$$\cos \alpha = \frac{h_1 h_2 + k_1 k_2 + l_1 l_2}{\sqrt{\sum h_1^2} \sqrt{\sum h_2^2}}$$

and the height d of the steps of $(h_1 k_1 l_1)$

$$d = \frac{a}{\sqrt{\sum h_1^2}}$$

where a is the lattice parameter. Simple geometry shows that the radius of the tip is

$$r = \frac{d}{1 - \cos \alpha} n$$

For a particular pair of planes and a particular metal $d/(1 - \cos \alpha)$ is a constant, i.e., $r = Kn$.

For the steps between (110) and (211) in tungsten

$$K = 16.68 \text{ Å per step}$$

Müller has published a useful table on page 150 of Ref. 13 which gives many of the essential constants needed for other pairs of planes, and Drechsler and Wolf[9] have published a detailed study of the variations of radii on a tungsten tip.

5.3 Spherical Surfaces of Solid Solutions

Experimentally it has been found that solid solution alloys of refractory metals give a highly distorted pattern in the field-ion microscope, and we propose to discuss here whether this unexpected result can be explained from a geometrical point of view.

The two pieces of experimental work which we shall consider are:

1. Ralph and Brandon,[10] who showed:
    a. An alloy of 5 At.% Re in W gave a field-ion microscope image with a regular structure and many planes visible that was similar to the image for pure W.
    b. An alloy of 25 At.% Re in W, on the other hand, gave a field-ion microscope image which was very irregular though only the (110), (112), and (100) planes could be distinguished.
2. Caspary and Krautz,[11] who showed:
    In the W–Mo series of alloys the effect on composition of the field-ion microscope pattern was not so marked as with W–Re, but the effect was nonsymmetrical with about 50% Mo and 50% W.

Thus any geometrical interpretation of alloy patterns should attempt to be compatible with the following experimental data:

1. A 5% alloy may give a normal pattern, but a 25% alloy may give a highly distorted one.
2. A pattern for 25% of B in A can be differently distorted than a pattern for 25% A in B.
3. The number of spots does not greatly alter for different alloys.

5.3.1   A Computer Program for Solid Solutions

The program previously used for spherical surfaces of pure metals can be adapted to alloys.

When P_0 was used as a critical distance for imaging, it was equivalent to assuming that during field evaporation protruding atoms were successively removed from the tip until an atomically smooth spherical surface was left and then atoms were imaged which were within a distance of 0 to P_0 from this.

One way we can consider a solid solution is to assume that the solvent atoms contribute to the image in the range of 0 to P_0 as before, but the solute atoms may evaporate more readily so that after field evaporation none would be present at a distance up to Q' below the surface. Also their critical distance for imaging would be P_0' and in general be different from P_0.

Thus the range over which the imaging solute atoms would be found is Q' to P_0'.

The data to the computer program therefore include the values P_0, Q' and P_0', the composition of the alloy in parts per 1000, the diameter of the tip, and the lattice parameter of the alloy.

In the program after the P value has been determined for each atom, a random number between 1 and 1000 is generated by the computer. If the random number is less than the composition of the alloy, the atom is assumed to be solute, and, if greater, it is solvent. The P value is then compared with the given values for solute or solvent, as the case may be, and the points plotted as an X or a 0 if P is within the appropriate range.

5.3.2 Matching with Experimental Patterns

We shall test for values which must be assumed for P_0, Q', and P_0' in order to obtain a distorted pattern similar to that made by Ralph and Brandon[10] for a tip of 25% Re–W alloy and about 200-Å radius. For this radius we have shown that, for a pure metal, a value of $P_0 = 0.1$ is appropriate, and this value shall be retained for the solvent (tungsten) in the 25% solid solution.

Let us assume that the solute atoms are field evaporated if they are less than 0.1 below the outer surface, but that they will contribute to the image if they are as much as 0.2 below it, i.e., $Q' = 0.1$ and $P_0' = 0.2$. The computed pattern (Fig. 5.12) shows more regularity than an experimental one, and it is clear that the properties of the solute have not been made sufficiently different from the solvent to give a disorganized pattern.

Let us assume therefore that both types of atoms are field evaporated together, i.e., $Q' = 0$, but solute atoms will image at a critical depth four times that of the solvent atoms, i.e., $P_0 = 0.1$ and $P_0' = 0.4$. The pattern (Fig. 5.12) is still more organized than an experimental pattern since planes can be distinguished in the (100)–(110) zone and several rings of (110) can be picked out.

Examples of this type, where the solute atoms relatively far below the surface show as image points, give a higher density of spots than are found by experiment. In this case there are 1.75 times the number of points given for pure solvent when $P = 0.1$. If the total number of points is adjusted by increasing Q' or proportionately decreasing P_0 and P_0', the pattern tends to become more regular, but we shall discuss this later.

We may also reduce the total number of image points if the solute atoms have been given the special property of preventing all neighboring solvent atoms from contributing to the image. (The program was adjusted to do this.) This would be equivalent to an electron shielding effect or to an interference with the supply of helium to the solvent atoms.

The pattern (Fig. 5.12) is disorganized like the experimental pattern.

Thus we must give the solute two special properties,

$$P_0' > 4P_0$$

and the ability to suppress the image from neighbors, if we are to match experimental patterns.

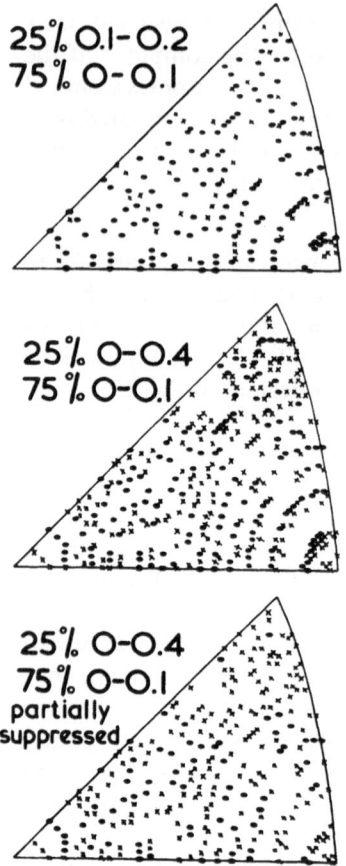

25% 0.1-0.2
75% 0-0.1

25% 0-0.4
75% 0-0.1

25% 0-0.4
75% 0-0.1
partially
suppressed

Fig. 5.12. Computer simulation of 25% Re–W alloy with 200-Å radius, with the use of various values of Q' and P'_0 and P_0 equal to 0.1 a_0.

5.3.3 The Solid Solution: A Lattice with Static Displacements

We can consider the criteria for imaging from a different point of view. We note the fact that the solid solution is made up of atoms of two different sizes, but, since the crystal structure is the same as that of the solvent, then all atoms must have their average positions statically displaced from their positions in an ideal lattice.

Webb[14] has proposed a simple formula relating the standard deviation of the static displacement with the concentration of solute and the limiting rate of change of the lattice parameter when the concentration of solute is at zero. Since every atom is displaced from its ideal position, then, in deciding whether such an atom will contribute to a field-ion image, its distance from the spherical surface must be compared with a value $p \pm \delta p$, which varies normally about p with a standard deviation of σ.

Computed patterns have shown that σ would have to be greater than 0.1 for the pattern to be disorganized like an experimental pattern. If the value of $\sigma = 0.15$ is put into Webb's formula, then a value of at least 1.03 Å is obtained for the rate of change of lattice parameter of W upon the addition of Re. Measured values for this are not known, but it is a much higher value than is known for any other W solid solution and 30 times higher than for the comparable system Mo–Re.[15-17] Thus it does not appear possible to explain images of the alloy structures in terms of static displacements.

5.3.4 The Asymmetry of Patterns about 50% Solute

The W–Mo alloys used by Caspary and Krautz[11] had tip sizes of 230 to 290 Å. With a pure metal of this size the appropriate number of image points is obtained if P_0 is about 0.08.

Suppose we allow P_0' to be four times P_0, as before, but we shall reduce the values of P_0' and P_0 so that for different compositions the number of image points remains constant at the value for pure metal.

Physically this could be interpreted that the field above solvent atoms is the same as that above solute atoms which are four times farther from the spherical surface. The actual number of atoms which image may be limited by some other factor, such as the supply of He to the tip.

The computed patterns for the three compositions investigated by Caspary and Krautz do show an asymmetry of disorganization around 50% solute, and the sense of the asymmetry indicates that we must assume that molybdenum is the solvent. That is, tungsten atoms will image more readily than molybdenum atoms.

Another method of obtaining patterns asymmetric about 50% solute is to assume that only solute atoms contribute to the image and that their limiting depth for image formation is adjusted for each composition to give a standard number of image points.

This is the condition which is approached if the above proportional reduction of P_0 and P_0' has been carried out with increasing ratios of P_0' to P_0.

Physically this can be interpreted in two ways:

1. The solute evaporates less readily than the solvent so that the surface layer of the tip is a skeleton of solute atoms.
2. Both solute and solvent evaporate at the same field and the surface has the proper proportion of solvent atoms, but the solute has the property of preventing their contributing to the image.

In Fig. 5.13 plots have been made for alloy compositions of 10%, 25%, 50%, and 75% with $P_0 = 0$ and $P_0' = 1.0, 0.4, 0.2$, and 0.133. A tip radius similar to that used by Ralph and Brandon was assumed.

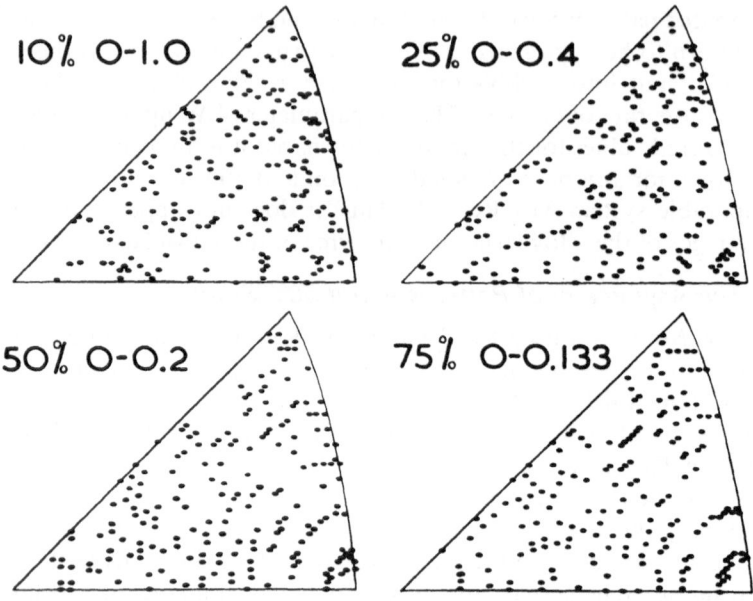

Fig. 5.13. Computer simulation of alloys with 10, 25, 50, and 75% solute. The P_0 is 0 and P_0' is 1.0, 0.4, 0.2, and 0.133, respectively, to maintain approximately constant density of image points.

The pattern for the 25% alloy is disorganized in a manner similar to their experimental pattern, and the asymmetry about 50% is apparent.

The organization of the pattern decreases with decreasing concentration of solute, and the 10% pattern was completely disorganized. This is contrary to the observation by Ralph and Brandon[10] that 5% W–Re alloys give a highly organized pattern.

However, in practice there would be a limiting concentration below which, depending on the properties attributing to the solute, either (1) the skeleton of solute atoms would become unstable and collapse and revert to its b.c.c. lattice, or (2) in order to maintain the number of image points, the image would have to come from solute atoms so far below the surface that solvent atoms much nearer the surface would no longer be prevented from contributing to the image.

Thus it appears that the asymmetry of the patterns can be reproduced by:

1. Allowing the solute atoms to be imaged at greater depths below the reference sphere than the solvent atoms.
2. Giving the solute atoms the property of reducing the depth for the imaging of solvent atoms below the value for pure solvent.

References

1. O. Knacke and I. N. Stranski, "The Mechanism of Evaporation," in: *Progress in Metal Physics*, Vol. 6, Pergamon Press (London and New York), 1956, pp. 181–235.
2. W. K. Burton, N. Cabrerra, and F. C. Frank, "The Growth of Crystals and the Equilibrium Structure of Their Surfaces," *Phil. Trans. Roy. Soc.*, London, **A243**: 299–358 (1951).
3. A. J. W. Moore and J. F. Nicholas, Atomic Configurations in Ideally Flat Surfaces; I—Construction of Models in Face-Centred and Body-Centred Cubic Crystals, *Phys. Chem. Solids* **20**: (1961), pp. 222–229.
4. J. F. Nicholas, "Atomic Configurations in Ideally Flat Surfaces; II—Description of Surface for an Arbitrary Crystal Structure," *Phys. Chem. Solids* **20**: 230–237 (1961).
5. J. K. McKenzie, A. J. W. Moore, and J. F. Nicholas, "Bonds Broken at Atomically Flat Crystal Surfaces; I—Face-Centred and Body-Centred Cubic Crystals," *Phys. Chem. Solids* **23**: 185–196 (1962).
6. J. K. McKenzie and J. F. Nicholas, "Bonds Broken at Atomically Flat Crystal Surfaces; II—Crystals containing Many Atoms in a Primitive Unit Cell," *Phys. Chem. Solids* **23**: 197–205 (1962).
7. J. F. Nicholas, *An Atlas of Models of Crystal Surfaces*, Gordon and Breach (New York), 1965.
8. E. W. Müller, "Beobachtung von nahezu fehlerfreien Metallkristallen und von Punktdefekten im Feldionenmikroskop," *Z. Physik* **156**: 399–410 (1959).
9. M. Drechsler and P. Wolf, Zur Analyse von Feldionenmikroskop-Aufnahmen mit atomarer Auflösung, *Intern. Conf. Electron. Microscopy*, 4th, *Berlin*, Springer Verlag, Berlin, 1958, pp. 835–848.
10. B. Ralph and W. G. Brandon, "A Field-Ion Microscope Study of some Tungsten–Rhenium Alloys," *Phil. Mag.* **8**: 919–934 (1963).
11. E. K. Caspary and E. Krautz, "Feldionenmikroskopische Untersuchungen im Mischkristallsystem Wolfram–Molybdän," *Z. Naturforsch.* **19a**: 593–595 (1965).
12. A. J. W. Moore, "The Structure of Atomically Smooth Spherical Surfaces," *Phys. Chem. Solids* **23**: 907–912 (1962).
13. E. W. Müller, "Field Ionization and Field-Ion Microscopy," *Advan. Electron. Electron Phys.* **13**: 83–179 (1960).
14. W. W. Webb, "Atomic Displacements in Metallic Solid Solutions," *J. Appl. Phys.* **33**: 3546–3552 (1962).
15. J. M. Dickinson and L. S. Richardson, "Constitution of Rhenium–Tungsten Alloy," *Trans. ASM* **51**: 758–771 (1959).
16. A. Taylor, N. J. Doyle, and B. J. Kagle, "The Ternary Alloy System Molybdenum–Rhenium–Hafnium," *Trans. ASM* **56**: 49–67 (1963).
17. W. B. Pearson, *Handbook of Lattice Spacings and Structures of Metals*, Pergamon Press (London and New York), 1958.

Chapter 6

ARTIFACTS, HYDROGEN PROMOTION, AND FIELD-ION MICROSCOPY OF NONREFRACTORY METALS

E. W. Müller†

6.1 Nonideal Crystals

Ideally the field-ion microscope specimen would be part of a sphere cut through the crystal lattice in a way that all atoms having their centers outside this sphere would be removed and all remaining surface atoms would be imaged. In reality the various physical effects connected with the surface preparation process, by field evaporation, and with the imaging process, by field ionization, cause deviations from the ideal situation. Many of them have been realized to exist since the early days of field-ion microscopy, but only recently has it been suggested to consider them as artifacts,[1] a term familiar to optical and electron microscopists. Sometimes these artifacts are hard to distinguish from intrinsic lattice imperfections, so that their study is of basic importance for the proper interpretation of field-ion micrographs.

In principle any image detail of a perfect crystal that does not correspond to the simple geometry of the idealized hemispherical cut through the undisturbed lattice, as, for instance, presented by Moore's computer model,[2] must be considered as an artifact (see also Chapter 5). In view of the crystallographic anisotropy of the field-evaporation end form there is actually no real metal that can be imaged without artifacts.

The crystallographic anisotropy caused by the evaporation process and elucidated by the theory of field evaporation[3–5] expresses itself in effects connected with the immediate vicinity of an individual atomic site as well as with the development of a specific regional tip radius. The latter is not of much concern since it has been shown by the author[6] how the local radius can be measured by counting the number of net plane rings between

† Research Professor of Physics, The Pennsylvania State University, University Park, Pa.

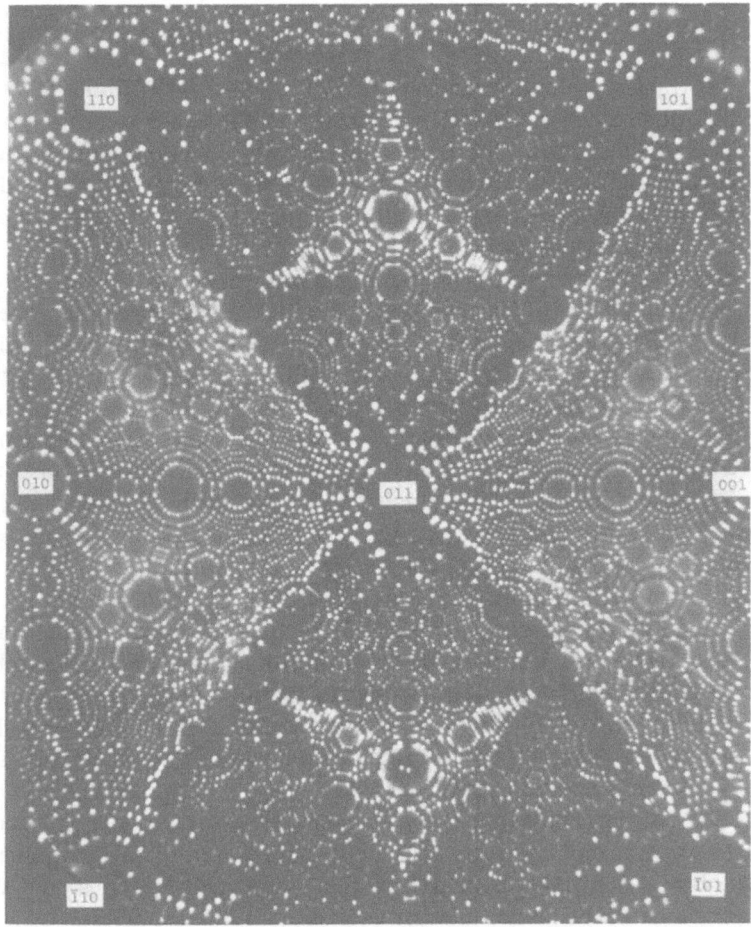

Fig. 6.1. Tantalum crystal showing regions of reduced ionization probability and metastable sites at one edge of {112} planes.

various crystallographic directions.† A more interesting effect, which does not disturb image interpretation but presents a problem worth further investigation, is the image brightness contrast between both sides of the [111] zone separating the square {001} regions from the {011} cornered triangles around the {111} planes. This effect is more pronounced on Mo and Ta (of which an example is shown in Fig. 6.1) than on W. On f.c.c. metals a very low ion current density is observed in the three narrow

† After the *5th Field Emission Symposium* in 1958 in Chicago, this method was adopted by Drechsler by his publishing a curvature map of a tungsten tip in the same *Proceedings*,[6] p. 835.

sections emerging from {111} and fading out smoothly toward the {233} regions. Even more characteristic for these metals is a sharply bounded, triangular bright region between the (011), (113), and ($\bar{1}$13) planes, centered near (012), and their symmetrical equivalents in the other crystallographic quadrants. Determination of local tip radius by net plane counting reveals no appreciable differences in local radii of curvature in the light and the darker regions. At some places the jump in brightness occurs right across the middle of one net plane.

Similar intensity changes within a narrow region are observed over the coherent boundaries of Guinier–Preston zones precipitated in a nickel–11 atomic % Be alloy.[7] Here the net plane rings crossing the boundary indicate again an equal radius of curvature on both sides, while the brightness contrast is of the order of magnitude 10 : 1. Regional field variation due to tip topography as well as orientation-dependent work functions cannot therefore be responsible for this effect. Rather a crystallographic anisotropy of field ionization seems to suggest itself. So far we have considered the tunneling probability of the image-gas electron into the metal as being only dependent upon the profile of the potential barrier. However, there may exist a sharply orientation dependent reflection coefficient for the electron about to tunnel out of the gas molecule.[7,8,25] High reflection co-efficients for electrons entering a metal with almost zero energy have been found experimentally,[9] and there seems to be no obvious reason why a reflection coefficient of considerable size could not exist at the Fermi level as well. That such a large reflection coefficient is there intrinsically rather than due to new surface states created by the strong external field is probably evidenced by a corresponding jump in the field-electron emission density of the field-evaporation end form at these regions. In ordinary field-emission patterns obtained from annealed tips this intensity step does not show because the high index planes are not well enough developed.

6.2   Metastable Sites

Before we can discuss other extended or regional artifacts, it is necessary to consider nonintrinsic defects of atomic dimensions, which have been puzzling us for a long time. There appear on the end form of all pure metals certain sites which cannot be obtained with the simple construction principles of the cork-ball or Moore models, such as the decoration of the [001] zone on tungsten or the [011] zone on platinum. There are other distinct but more randomly arranged sites in a region between the (253) and (132) planes of tungsten and even more pronounced in the same region on tantalum. These specific sites which were for a while considered as possible impurities, have recently been explained[10] as low-coordination surface sites which have been stabilized by the evaporation field. Stranski's[11] old idea of summing up the binding energy of a surface atom from the contributions of nearest and next nearest neighbors, although hard to reconcile with the electron theory of the metallic bond, has in practice been very useful for estimating binding energies at various surface sites. Under high-field conditions the increase of polarization binding energy due to the more exposed position of an atom with a smaller coordination number may become larger than the loss of binding energy by the reduction of coordination. On a b.c.c. lattice, for instance, atoms on a kink site from which evaporation normally proceeds have four nearest neighbors, each roughly contributing one-fourth of the sublimation energy Λ. Its atomic polarizability at that site may be designated by α_{a4}. Field stabilization at a site with a coordination number of 3 will occur when the additional binding energy $\frac{1}{2}(\alpha_{a3} - \alpha_{a4})F^2 > \Lambda/4$.†

For obvious reasons we cannot presently calculate the atomic polarizabilities of surface atoms; however, this polarizability should approach that of a free atom as the coordination number decreases. In the case of tungsten with $\Lambda = 8.67$ eV at a field of 560 MV/cm, $\alpha_{a3} - \alpha_{a4}$ only needs to be larger than 2×10^{-24} cm^3, a plausible value when compared with the polarizability of a free tungsten atom $\alpha_0 = 7 \times 10^{-24}$ cm^3 as measured by Liepack and Drechsler.[13] The delicacy of balance between an ordinary 4-coordination kink site and a field-stabilized 3-coordination site becomes apparent when the temperature dependence of the appearance of these sites on tungsten at the edge of the [111] zone is considered. Here the field-stabilized sites appear only when field evaporation is performed well below liquid-nitrogen temperature, so that one can immediately judge whether a tungsten crystal has been evaporated at liquid-nitrogen or liquid-hydrogen temperature.[10]

The [001] zone decoration by metastable atoms is not very temperature sensitive on tungsten, but these sites are not a general feature of the b.c.c. lattice. We find it weakly developed on iron if the tip radius is large enough but not on molybdenum nor on niobium. On tantalum a similar decoration appears at the [0$\bar{1}$1] zone when the evaporation is performed above room temperature. On this metal we find the metastable sites on the one side of the [111] zone, which were first seen on tungsten, much more pronounced and extended over a larger area (Fig. 6.1). The removal of this field-stabilized "disorder" requires field evaporation above about 150°K.

On f.c.c. platinum and iridium there is a very distinct field-stabilized decoration of the [110] zone. The temperature range of its appearance is not yet clearly established; it seems that evaporation above about 50°K is essential. So far we have not observed any distinct zone decorations on copper, gold, palladium, rhodium, nickel, or f.c.c. cobalt.

† The contribution of next nearest neighbors[12] could be included easily, but is unnecessary in view of the uncertainty in the α_a values.

The chemical specificity of the zone decoration is also evident for the few examples of hexagonal close-packed (h.c.p) metals that are available. Rhenium shows extremely well-developed [11$\bar{2}$1] zone decorations, particularly when a crystal with the c axis parallel to the tip axis is available. Hexagonal close-packed cobalt when field evaporated with hydrogen promotion does not develop these decorations, and none have been seen on zirconium either. On most of the metals mentioned so far the field stabilization of low-coordination lattice sites is not very disturbing for the observation of intrinsic lattice defects because the metastable sites are not numerous and are limited to narrow crystallographic regions. This is not quite so for tantalum, where the disordered regions near the {112} planes occupy between 5 and 10% of the total image area. With niobium the situation is very bad as metastable sites occur over wide areas, almost over the entire crystal hemisphere.

Although we are presently not concerned with the process of image formation, it should be pointed out that a probable explanation of the great brightness of the atoms in metastable sites lies in the increased field penetration[10] which is also responsible for the increased binding energy. Field penetration reduces the critical distance of the ionization zone from the tangible surface, so that an additional supply of low hopping atoms becomes available for ionization.

Unfortunately the appearance of the metastable atom spots is not different from impurity spots due to oxygen or some other interstitials or even adsorbed atoms. The abundance of metastable sites on niobium makes this metal particularly very unsuitable for the study of interstitials, which can be easily introduced into this metal by diffusion. When field evaporation of niobium is performed at or above liquid-nitrogen temperature, the metastable sites no more appear and interstitials become more easily visible as fairly large blobs. Because of the high evaporation rate at this temperature and helium imaging field, the effects are transient and require cinematographic or single-shot photography through an image intensifier. This change of the niobium surface from a random appearance due to metastable sites at 21°K to the flashing up of transient interstitials at 78°K is shown in a film.

6.3   Artifact Vacancies

Perhaps even more disturbing than metastable sites are artifact vacancies. These may occur in a variety of situations. Beginning with an almost trivial case, vacancies can be suggested by burnt-in dead spots of the phosphor screen. Such sites can be recognized during visual observation as stationary spots by wiggling the image with a magnet, by going to a higher field so that the screen is more uniformly illuminated, or by proceeding with field evaporation to change the surface. If only one photograph is available, dead-screen "vacancies" can be recognized by their unusual contrast and their crystallographic location at places where bright atom spots most likely have been situated in a preceding image.†

True surface vacancies are produced as artifacts by various mechanisms. While, in a pure metal, field evaporation always proceeds via kink sites or high-field corners of net plane edges, point defects in the next lower layer may alter this sequence. If there is an intrinsic vacancy below the surface, the covering atom in the upper layer may drop into it, particularly when the recessing layer edge approaches the site. Thereby the net plane becomes indented, as has been observed in α-irradiated material.[15] While this in effect only shifts the location of a vacancy, an underlying impurity or interstitial atom may reduce the binding energy of the matrix atom above so much that it will field evaporate from the middle of a net plane, as was observed by Machlin.[16] In dilute alloys preferred evaporation of the solute may again result in the formation of surface vacancies either at net plane edges or within the plane.

Field evaporation in vacuum as well as in the presence of pure helium or pure neon always proceeds by removing atoms from net plane edges. However in a helium–neon gas mixture [17] near the onset of field evaporation many vacancies are produced inside fully resolved net planes in the low-work function regions. Even if the neon partial pressure is a small fraction of the helium gas pressure, this process still goes on. These vacancies are produced by the impact of helium atoms having the dipole attraction energy $1/2\alpha F^2$ and being capable of transferring this energy efficiently to the metal atom when the temporarily adsorbed neon atom acts as an intermediate collision partner.

The most frequent cause of a surface vacancy seems to be field-induced chemical etching.[18,19] An abundance of surface vacancies has been demonstrated to appear when adsorbates are removed by a desorption field.[20] In the case of W, Ir, and Pt, atomic nitrogen leaves the surface by taking a metal atom with it, while the desorption of a molecule does not attack the metal substrate (Fig. 6.2). These surface vacancies are formed with almost equal ease in the middle of a net plane or on the edge. A direct mass spectroscopic proof of the field evaporation of molecular species has been obtained so far only in the case of hydrogen and deuterium reacting with copper,[18] where the ion species CuH_2^+ and CuD_2^+ were found.[21]

The attack on pure metal surfaces by hydrogen and deuterium is limited to the normal field-evaporation sites at the edge of net planes. Impurity atoms, however, may be picked out from the middle of net planes. The abundance of surface vacancies on deuterium-promoted niobium[22] is

† See, for instance, Fig. 11 in Ref. 14.

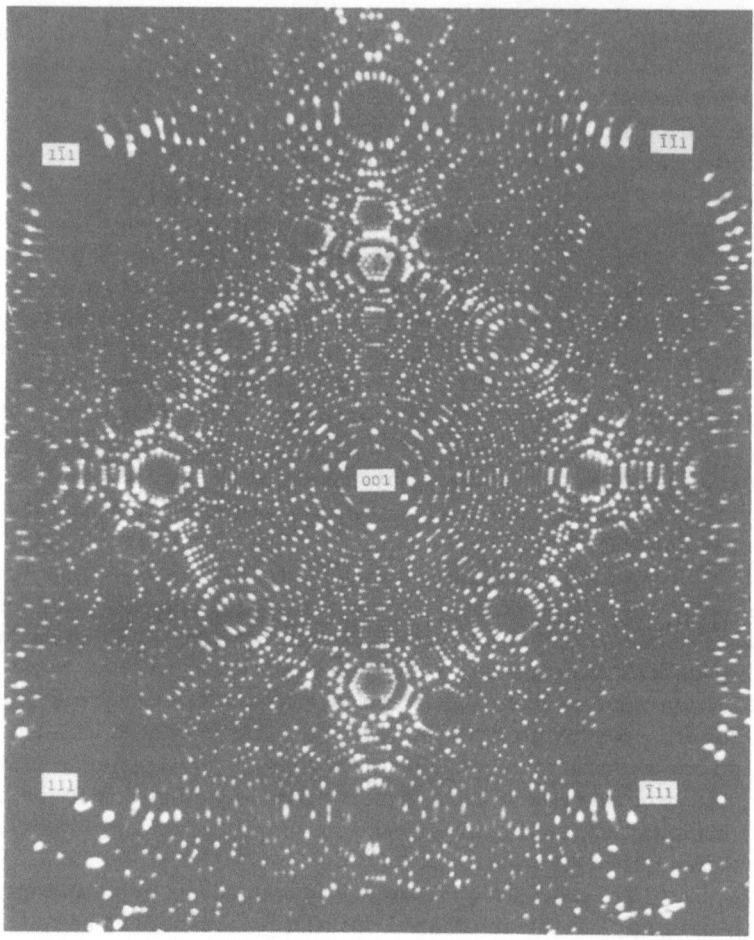

Fig. 6.2. Platinum tip with artifact vacancies caused by desorption of nitrogen.

a representative example (Fig. 6.3a and b). It appears that the resulting surface-vacancy density corresponds closely to the concentration of impurity interstitials in that material, probably oxygen and nitrogen.

The discussion of point defects as artifacts should be concluded by mentioning the possibility of radiation damage occurring when negative ions are attracted to the tip and impinge with the energy of the full acceleration voltage. Such negative ions can be released from the accelerating electrode when it is hit by the image-gas ions. This may happen when the emitted conical ion beam does not conform to the opening in the electrode. The efficiency of negative-ion release by impact of helium ions is low, so

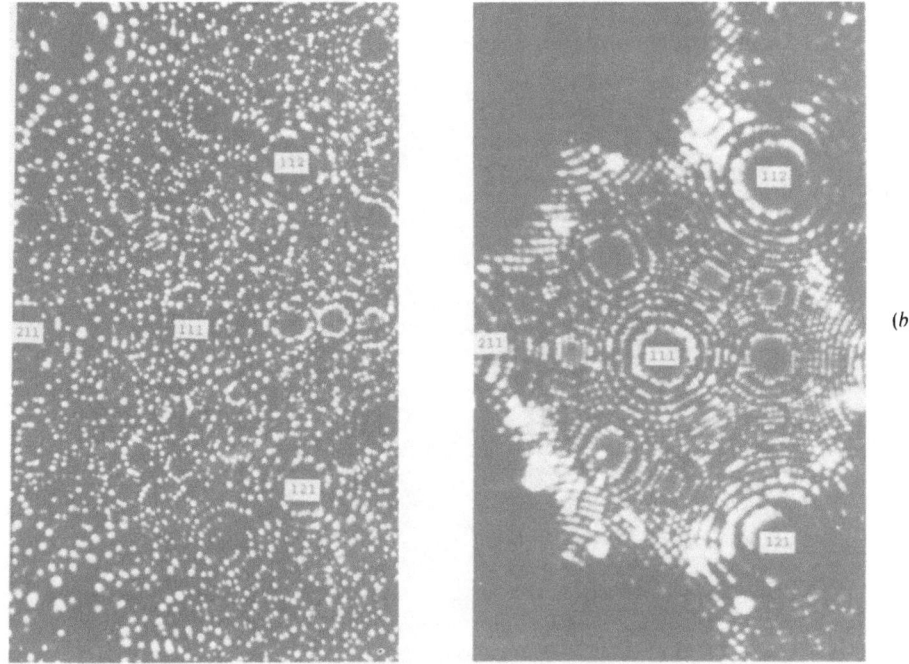

(a)   (b)

Fig. 6.3. (a) Section of niobium tip field evaporated at 21°K in He, having many Nb atoms at metastable sites. (b) Same specimen field-evaporated and imaged at reduced stress with He–D_2 mixture. Surface vacancies are formed by reaction of deuterium with oxygen interstitials.

that this effect was discovered only when very high ion current densities resulting from an operational gas pressure of 60 to 80 mtorr was used in a dynamic gas supply microscope.[23] Negative-ion release was found to be occurring abundantly with aluminum and stainless steel electrodes, and the trouble could be cured by gold-plating the accelerator electrode. Bombardment damage artifacts occur more frequently and at ordinary gas pressure when the imaging gas is neon or a neon–helium mixture because the heavy neon ions released negative ions more efficiently.[17]

6.4 Extended Artifacts

The tremendous mechanical stress $F^2/8\pi$ exerted by the electric field at the tip surface has been a matter of concern since the inception of the microscope.[24] At 500 MV/cm it equals about 1 ton/mm². This stress may exceed the technical yield strength of tip materials by one or two orders of magnitude. The shear component of the stress, due to the nonsphericity of the specimen, may cause dislocations to move, sometimes until the tip fractures.

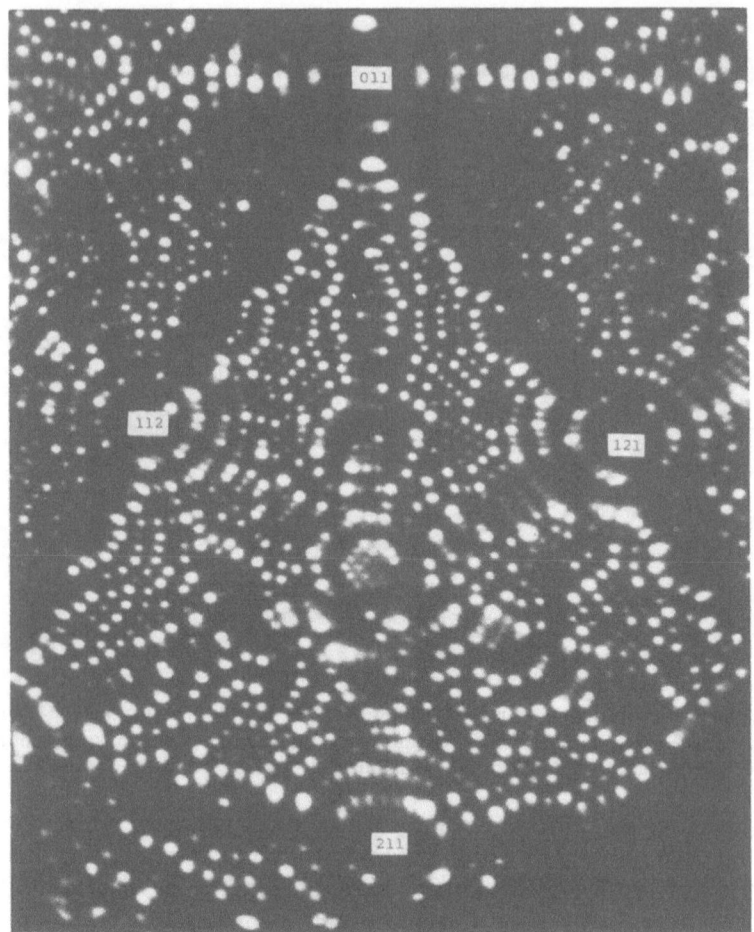

Fig. 6.4. Intersection of dislocation with the (111) plane and other defects formed after yielding of molybdenum tip due to field stress.

Yield in restricted areas under the field stress is typical for even the refractory molybdenum. A close network of dislocations emerges in the (111) regions, while other areas of the crystal hemisphere remain perfect (Fig. 6.4). It appears that only with a tip of very small radius, below 200 Å, is there insufficient room for the defects to grow. Niobium behaves very similarly, while tantalum and tungsten are strong enough not to yield at their respective evaporation fields. Of the other b.c.c. metals, only iron has been investigated thoroughly. Reflecting the strong anisotropy of Young's modulus of iron, the (111) region, specifically, assumes a completely amorphous structure. As the number of next nearest atomic neighbors is

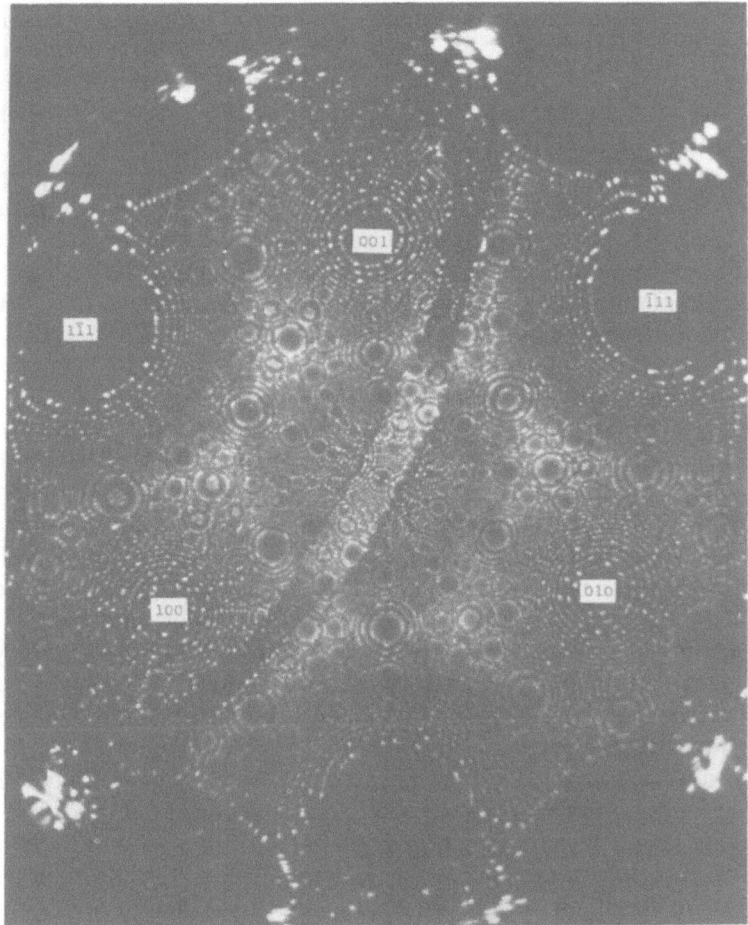

Fig. 6.5. Iridium specimen with a twin "slice" formed spontaneously by the field stress.

diminished, field evaporation on the (111) region occurs so readily that a depression of the surface results, into which one can "look" only by using neon as the imaging gas. Of the metals with f.c.c. lattices, iridium, platinum, rhodium, and, most surprisingly, also gold, and perhaps copper do not yield under their evaporation field, except that iridium tips in {111} orientation tend to develop twin slabs (Fig. 6.5). Palladium and particularly nickel, on the other hand, develop dislocation networks in the (102) and (113) regions, while both the (001) and the (111) regions remain intact.

It is doubtful whether the field-induced dislocation tangles grow from intrinsic defects. Rather, nucleation of dislocation loops seems to take place. The field stress may be considered a negative hydrostatic pressure of approximately 100 kilobars at 500 MV/cm. Under this condition the ordinary bulk modulus of the metal, as known from compression measurements, suggests a volume expansion of some 5 to 10%. Actually, the elastic bulk modulus at a large negative pressure, so far not accessible to measurement, must be even larger because of the asymmetry of the atomic potential function. Elastic volume expansions exceeding those connected with the thermal expansion to the melting point must occur, and spontaneous vacancy formation appears possible.[25] Let us take, as an example, the volume V_a of a vacancy in platinum to be equal to the atomic volume 15.1×10^{-24} cm^3; the energy of increasing the volume of the tip is

$$p \, dV = \frac{F^2}{8\pi} V_a = 0.94 \, \text{eV}$$

at the evaporation field of 475 MV/cm. This energy is 20% less than the actual energy of formation of a vacancy as measured from quenching experiments or by direct counting of vacancies with the field-ion microscope, so that in the case of platinum we are assured of the absence of artificial formation of vacancies by the evaporation-field stress. This is obviously not true for metals such as nickel or iron; but, unfortunately, for most metals neither the volume nor the energy of formation of a vacancy are known accurately enough to predict the maximum field stress that could be applied without vacancy formation and a subsequent collapse of vacancy disks to form dislocations.

It is a fallacy to assume that all these field-stress-induced artifacts can be prevented by operating the microscope at a reduced field strength with an imaging gas of lower ionization potential. In order to make the radial projection principle of the microscope function properly, it is first necessary to shape the specimen surface by field evaporation before attempting to image it. We can now formulate the basic rule[1,10] for the applicability of field-ion microscopy to various metals without undue interference by artifacts: Field evaporation must be performed below the field that causes the metal to yield. The use of low-ionization-potential gases for imaging is only of secondary importance.

6.5   Reduction of the Operating Field by Hydrogen

Field evaporation can be performed at a reduced field by adding a small amount of hydrogen or deuterium to the imaging gas. The effect of hydrogen promotion was discovered a few years ago[19] but its exact nature is still obscure.

Upon the addition of hydrogen or deuterium the field evaporation rate increases approximately proportional with the partial pressure, but to a different degree with various metals. The evaporation of copper and iron is very much accelerated, while for rhodium the increase in rate is small. A very useful feature of hydrogen promotion of an already field-evaporated specimen of W, Mo, or Ta is that further removal of atoms occurs preferentially at the more protruding {111} and {001} tip regions, so that the final form is more evenly curved and thereby gives an improved ion image. While for these metals the evaporation rate in the presence of hydrogen goes down when the applied voltage is reduced, a quite different behavior is observed with platinum, nickel, and iron. These metals show a slow reaction with hydrogen at helium BIV, but the evaporation rate increases considerably when the tip voltage is reduced to 70% of helium BIV. This is obviously due to the gradually disappearing shielding effect of the high field around the tip, which hinders the approach of the easily ionizable hydrogen molecules. At this lower evaporation field the stress is reduced to only one-half of the previously effective field stress and is now well below the yield stress of the tip materials.

During the study of the evaporation in the presence of hydrogen a quite surprising and welcome bonus was obtained when it turned out that the field strength required for ionization of helium was also considerably reduced.[22] This can be understood by a close consideration of the imaging process connected with the hopping motion of the gas atoms. In order to obtain a good resolution, the tangential velocity component of the hopping helium atom must be small.[26] This is ordinarily achieved by a gradual reduction of the large kinetic energy of the impinging image-gas molecules in a sequence of hops. At each collision with the surface the temperature of the gas is reduced according to the effective accommodation coefficient, which is only of the order of 0.02 to 0.01 for helium on tungsten. Thus it will take some 250 hops to approach the tip temperature of $21°K$ in the typical case of liquid-hydrogen cooling. A good fraction of helium atoms is actually ionized before full accommodation has been achieved. This requires a relatively high field because of the short transit time in the ionization zone, and the resulting ions still have a fairly large tangential velocity component. The presence of hydrogen, probably in an adsorbed state of the surface, provides an intermediate collision partner of low mass with which the light helium atoms can exchange their energy quite effectively. The number of hops for full accommodation to the low tip temperature is greatly reduced, and the population of slowly hopping helium atoms increased. Now ionization can be achieved at 70% of the previous image field, and the resolution is better than before. The adsorbed hydrogen atoms that produce this dramatic effect are not visible in the ion image. It is possible that the hydrogen is adsorbed as an atom on recessed sites between the surface atoms, or it may be present at the shank only, from where a large supply of hopping atoms comes. In order to image metals such as iron, nickel, and cobalt, the partial pressure of hydrogen must be carefully monitored, for instance with a zirconium getter.[27] Iron is particularly sensitive to hydrogen. After promoting field evaporation at a partial pressure of 10^{-6} torr the gas is removed by a titanium getter, and a subsequent field evaporation of one or two more surface layers will then produce a very regular surface.

With the introduction of this new hydrogen promotion technique the previously nearly inaccessible first series of the transition elements, particularly iron, nickel, and cobalt have come within reach of the field-ion microscope. It appears that the introduction of the image intensifier,[28] the recognition of the significance of artifacts,[10] and the use of hydrogen promotion[22] for field evaporation and imaging mark some of the most significant advances in the practice of field-ion microscopy since the establishment of the standard observational procedures[19] before 1960.

References

1. E. W. Müller, Xerox–Cornell Materials Science Center Lecture, March, 1964. Materials Science Center, Cornell University, Report # 276.
2. A. J. W. Moore, *Phys. Chem. Solids*, **23**: 907 (1962).
3. E. W. Müller, *Phys. Rev.* **102**: 618 (1956); also *Advan. Electron. Electron Phys.* **13**: 83 (1960).
4. R. Gomer and L. Swanson, *J. Chem. Phys.* **38**: 1613 (1963).
5. D. G. Brandon, *Surface Science*, 3: 1 (1965).
6. E. W. Müller, *5th Field-Emission Symposium*, Chicago, June 1958, also: *Proc. 4th Intern. Conf. Electron Microscopy, Berlin*, Aug., 1958, Vol. 1, Springer Verlag (Berlin), 1960, p. 820.
7. S. B. McLane and E. W. Müller, *12th Field-Emission Symposium*, Pennsylvania State University, University Park, Pa., 1965.
8. S. Nakamura and E. W. Müller, *J. Appl. Phys.* **36**: 2535, (1965).
9. R. J. Zollweg, *Surface Science* 2: 409 (1964).
10. E. W. Müller, *Surface Science* 2: 484 (1964).
11. I. N. Stranski, *Z. Physik. Chem. B.* **11**: 342 (1931).
12. E. W. Müller, *Z. Physik* **126**: 642 (1949).
13. H. Liepack and M. Drechsler, *Naturwiss.* **43**: 52 (1956).
14. D. G. Brandon and M. Wald, *Phil. Mag.* **6**: 1035 (1961).
15. E. W. Müller, in: *Imperfections in Crystals*, J. B. Newkirk and J. H. Wernick, eds., Wiley (Interscience) (New York), 1961, p. 77.
16. E. S. Machlin and W. DuBroff, *12th Field-Emission Symposium*, Pennsylvania State University, University Park, Pa., 1965.
17. O. Nishikawa and E. W. Müller, *J. Appl. Phys.* **35**: 2806 (1964).
18. E. W. Müller, in: *Structure and Properties of Thin Films*, C. A. Neugebauer, J. D. Newkirk, and D. A. Vermilyea, eds., Wiley (New York), 1959, p. 476.
19. E. W. Müller, *Advan. Electron. Electron Phys.* **13**: 83 (1960).
20. J. F. Mulson and E. W. Müller, *J. Chem. Phys.* **38**: 2615 (1963).
21. E. W. Müller and Rebne Thomsen, unpublished.
22. E. W. Müller, S. Nakamura, O. Nishikawa, and S. B. McLane, *J. Appl. Phys.* **36**: 2496 (1965).
23. B. J. Waclawski and E. W. Müller, *J. Appl. Phys.* **32**: 1472 (1961).
24. E. W. Müller, *Z. Physik* **131**: 136 (1951).

25. E. W. Müller, *Science* **149**: 591 (1965).
26. E. W. Müller, *J. Appl. Phys.* **27**: 474 (1956) and **28**: 1 (1957).
27. E. W. Müller, S. Nakamura, S. B. McLane, and O. Nishikawa, *Proc. 3rd Intern. Vacuum Congr.*, Stuttgart, 1965, Vol. 2, p. 431.
28. S. B. McLane, E. W. Müller, and O. Nishikawa, *Rev. Sci. Instr.* **35**: 1297 (1964).

Chapter 7

INTERPRETATION OF FIELD-ION MICROSCOPE IMAGES OF POINT AND LINE DEFECTS

J. J. Hren†

7.1 Introduction

It has been recognized for some time now that field-ion microscopy offers a unique possibility for examining crystalline defects at the atomic level. To some degree this potential has been realized, but progress has been mixed to date.

To be sure, images purporting to represent defects of nearly every kind have been reported (see, e.g., Ref. 1). In the case of point defects some quantitative studies have been made, and most of the numbers seem to agree with expected values. Although no strikingly new or unexpected observations on point defects have as yet been reported, the technique seems to be on the verge of yielding important data. The distribution of point defects already appears to be a tractable study to conduct with field-ion microscopy, especially as related to radiation damage (*cf.* Chapter 11). Even more promise appears to lie in studies of secondary defects (e.g., dislocation loops, stacking faults, and tetrahedra) resulting from recovery processes following quenching, irradiation, or deformation.

Experimental information concerning line defects has, up until very recently, been qualitative or speculative only, and the hoped for ultimate answers or atomic analyses of, say, dislocation core structures have not yet emerged from field-ion microscope (FIM) studies. Some very encouraging results have been reported of late,[2,3] but it is worthwhile, nevertheless, to pose the question: Is such information indeed retrievable at all from FIM images? Let us explore answers to this query by reviewing some of the recognized experimental variables before describing the progress made to date.

† Department of Metallurgical and Materials Engineering, University of Florida, Gainesville, Fla.

7.2   Undissociated Dislocations in Field-Ion Microscope Specimens

A logical starting point in this analysis is with the intrinsic properties of line defects. We shall restrict ourselves exclusively here to undissociated dislocations, since stacking faults will be treated in Chapter 8. What are some of the pertinent properties of dislocations?

7.2.1   Geometry

A primary parameter of interest to field-ion microscopists is the geometry of a defect. It is this property which manifests itself in a FIM image and from which a researcher would hope to verify, refute, or extend present theories on the strength of crystals. There are a number of complications, as we shall now see.

To begin with, the atomic geometry of all but the simplest dislocations, i.e., pure edge and pure screw, are complex in an atomic ball lattice of the most symmetric crystalline structures. To prove this, try visualizing even the simplest nonperpendicular cut through a dislocation line. A quick perusal of Read's classic drawing[4] will serve to reinforce this opinion. Figure 7.1 is a

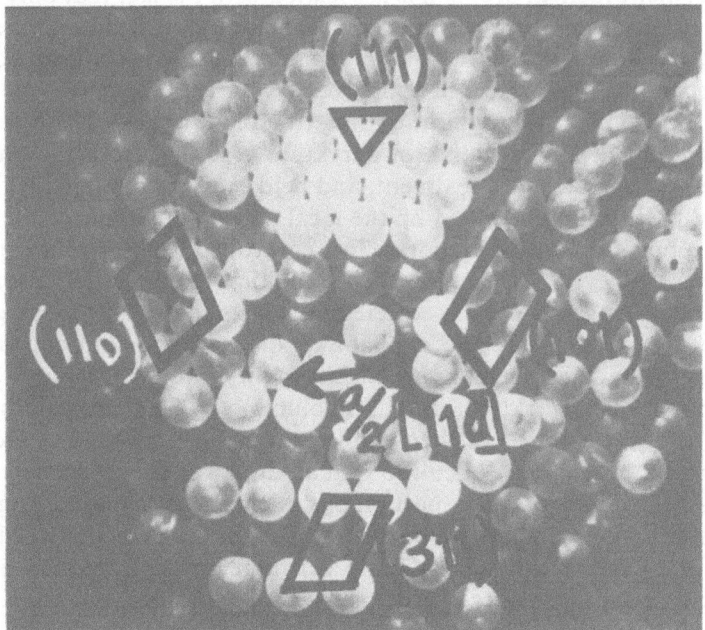

Fig. 7.1. Ball model showing core structure of dislocation with Burgers vector $a_0/2$ [110] intersecting the (311) plane.

further example. It shows a ball model of the intersection of an edge dislocation with $\mathbf{b} = a_0/2 \,[110]$ in an f.c.c. lattice intersecting the (311) plane. One needs now to extend such an analysis to the intersection of a dislocation of arbitrary character with any crystallographic plane. This is clearly a straightforward yet complicated problem, one that might be well suited for analysis with the aid of a computer as an extension of the kind of work described in Chapter 5. The atomic geometry in the vicinity of a dislocation core is obviously difficult to visualize except in the most straightforward case.

It should be noted that an analysis of the intersection of a dislocation line and a plane is for practical purposes the only pure geometrical problem one needs to solve. This is so since the approximately hemispherical specimen surface is in reality comprised of a series of facets corresponding to the various crystallographic planes.

The geometrically long-range effect (over some tens or a hundred atom spacings), on the other hand, has proven to be considerably more amenable to study. That is, the evidence of a disturbance in atomic packing may be detected in many cases over a rather wide area of the image (compared with the dislocation core). We shall discuss some recent results in this area in Section 7.4.

A further practical problem is associated with the fact that only a small fraction of the atoms visible on the surface actually contribute to the image at any given time (see Chapter 5). Some planes may give an image of every atom, depending on the facet size and average radius; others will give images only of the atoms at the outermost ring (the plane edges). The image thus depends upon the local radius of curvature resulting from the end form produced by preparation, treatment, and field evaporation (mostly, but not exclusively the last named) and on the local density of atomic packing. It is, of course, possible to field evaporate in a controlled way such that an image is obtained of every atomic site. This requires great patience, excellent electronics, a long time, and a coterie of analysts to peruse the images that are photographed. Since the probability of even encountering a defect (especially a line defect) in many specimens is low, it would be highly desirable to be able to scan quickly "through a volume" (by field evaporation) until a defect is encountered and then employ atom by atom field evaporation. Unfortunately, it is quite difficult to spot all but gross defects while imaging; it generally requires careful inspection of a micrograph. Thus one might have to make hundreds of micrographs first and examine them afterwards in the hope of finding some defects to analyze. The recording step may be accelerated by image intensification of one kind or another, but the analyst's time is still the limiting factor at present. Where there is a fairly high density of defects, such as point defects and clusters following substantial irradiation, this is not too distasteful a procedure; however for most studies this experimental technique is quite analogous to bubble chamber analyses—necessary, since it is the only practical way at the moment, but hardly something to look forward to.

In any case, it should be clear that the *core* geometry of a dislocation, already difficult to visualize when all the lattice points are visible, becomes even less amenable to study because of the practical difficulties of: (a) not generally being able to detect it in the image before recording and (2) only being able to image a fraction of the atoms about it at any given time. In addition, the local geometry may be badly distorted in the image owing to the local perturbations in the interatomic potentials caused by the defect. Streaking of the image e.g., can occur (see Section 8.5).

Besides geometry, there are other factors that must be taken into account.

7.2.2 Stresses Present in Specimen

It has been stated many times that the stresses present in a FIM specimen under imaging conditions are very great. According to classical electrostatics these stresses are of the order of

$$\sigma = \frac{F_0^2}{8\pi} \tag{7.1}$$

where F_0 is the field strength. The field varies as

$$F = F_0 \left(\frac{R}{r}\right)^2 \tag{7.2}$$

where R is the specimen radius and r is some distance such that $r > R$. If the tip is a hemisphere, the stresses are thought to be nearly hydrostatic over a solid angle of about $3\pi/2$, tapering off sharply down the shank (as shown in Fig. 4.6). In this transition region where the hemispherical cap joins the conical shank, a significant stress gradient is acknowledged to exist. Without doubt large shear stresses are also present. These latter can often lead to massive plastic deformation and indeed destruction of the specimen (known as *flashing*). If the specimen cross section is not symmetric, further shear stresses may be developed throughout the specimen. A still further complication arises because the tip is not truly hemispherical but consists of facets of varying size. The relative size of these facets is roughly proportional to the inverse density of atomic packing on the plane,[5] but local anisotropy of the inner potential and interatomic binding can modify this considerably. In any event, the specimen end form (following field evaporation) may deviate considerably from a hemisphere over even small solid angles. Figure 7.2 after Müller[6] illustrates this feature vividly. Drechsler and Wolf[7] have also treated the surface geometry of tungsten at some length. They

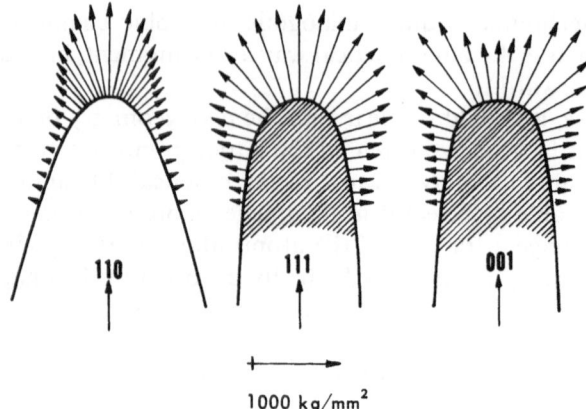

Fig. 7.2. Illustrating several possible end-form geometries and the variation in surface forces resulting from the applied electric field. (After Müller.[6])

found differences in local radii of more than a factor of 2. An application of their analysis to any field-ion image will almost invariably yield similar differences in radii from region to region. At best, then, one has a surface comprised of radii differing by at least a factor of 2 from region to region. According to equations (7.1) and (7.2) the stress should correspondingly vary by something like a factor of 8. It is obviously not possible to dismiss these high stress gradients as principally hydrostatic and hence not seriously affecting most defects present in the specimen. Unfortunately, a detailed treatment of the stress distribution and magnitude has not yet been reported or even, to the author's knowledge, been seriously attempted to date. Without this knowledge there is much uncertainty concerning any new information about defects that may come from field-ion microscopy.

7.2.3 Effects of Specimen Size and Imaging Stresses

Dislocations. A FIM tip is, in effect, a very fine whisker and (except for very high intrinsic dislocation densities) the probability of finding a dislocation within a hemispherical cap is usually small to begin with. For example, with a dislocation density $\rho = 10^{12}$ cm/cm^3 (uniformly distributed) and a tip radius of 1000 Å, one would expect between 10 and 100 dislocations to intersect the surface. However for $\rho = 10^9$ cm/cm^3 the probability of a dislocation intersecting the surface is 1 in 10 or less. On the other hand, for dislocation densities of 10^{10} cm/cm^3 or greater (where there is a reasonable probability of at least finding a dislocation in the image) there are considerable interactions between individual dislocations and groups of dislocations, resulting in high internal stress fields. Work-hardening theory of pure metals suggests that long-range stress fields extending to several microns may very likely exist. In preparing a specimen by electrochemical or chemical etching these internal stress fields are relieved as the dislocations slip out of the crystal. This is known to occur when thin foils are prepared for transmission electron microscopy and there is no reason to doubt that it will occur during preparation of FIM specimens.

There are also several mechanisms by which dislocations can leave the crystal before the imaging voltage is attained.

1. Significant shear stresses are developed as the electropolished specimen is taken up to the field-evaporation voltage for the first time. That is, the specimen at this stage is rather rough on an atomic scale. Thus, as the applied potential is increased, stresses build up most rapidly in areas with the smallest local radius of curvature. These then field-evaporate first, which leaves a larger radius, etc., until the entire specimen has been field evaporated. In this process there are large local variations in the stress fields and significant shear stresses are undoubtedly developed. These latter once again favor the annihilation of dislocations as they slip out of the specimen. One might expect the resulting slip steps to be unstable and field-evaporate; however Müller[8] has shown micrographs of stable slip bands in platinum deformed in the microscope. As an aside, this is strong evidence that the field-evaporated end form is not a sacrosanct geometric shape, since these slip bands give a stable image over many atomic layers of field evaporation.

2. The dimensions of the crystal in the volume imaged are sufficiently small that dislocation image forces (resulting from the proximity to the surface and not to be confused with the stresses developed when the specimen is imaged) will at least tend to rearrange any dislocations present. There is also a strong possibility that at least off-axis dislocations (those not substantially colinear with the wire axis) will either slip out of the surface or be effectively removed from the imaged region.

It is informative to consider the probability of a dislocation being "pulled" out of the crystal by the image stress somewhat more quantitatively. To do this we use the results of Bullough[9] concerning the interaction of dislocations with a free surface. Bullough considers both pure edge and pure screw dislocation aligned parallel to the axis of a cylinder (see Fig. 7.3). The result for pure screw is

$$\mathbf{P_s} = -\frac{\mathbf{b}\mu}{2\pi} \frac{r - \xi}{\xi(2r - \xi)} \tag{7.3}$$

where μ is the shear modulus, $\mathbf{b}$ the Burgers vector, and r and ξ are defined in Fig. 7.3.

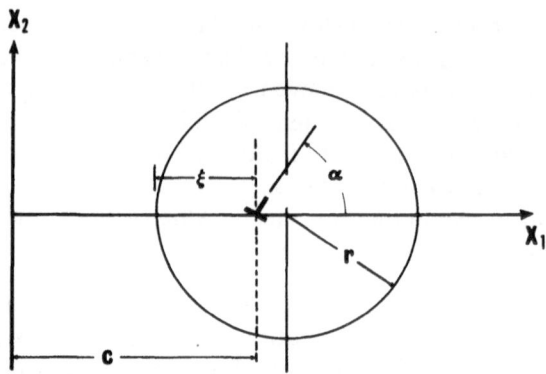

Fig. 7.3. Definition of parameters used to calculate image force of a screw dislocation in idealized field-ion specimen. (After Bullough.[8])

We can now evaluate $\mathbf{P_s}$ as a function of ξ for suitable values of r, $\mathbf{b}$, and μ. For pure screw (the only case we shall treat here)

$$\mathbf{P_s} \to 0 \qquad \text{as} \qquad \xi \to r \tag{7.4a}$$

and

$$\mathbf{P_s} \to \infty \qquad \text{as} \qquad \xi \to 0 \quad \text{or} \quad 2r \tag{7.4b}$$

This says simply that a pure screw dislocation is stable in the center of a cylinder and progressively less stable as it is moved off axis. It is more meaningful, however, to compare $\mathbf{P_s}$ with the intrinsic stress required to move a dislocation (i.e., the Peierl's stress) in a particular material. This is difficult to deduce with any accuracy; however, as an estimate we can use some fraction (say $\frac{1}{100}$) of the shear modulus of the slip plane in the direction of the Burgers vector. Table 1, taken from Reid[10] gives the pertinent elastic constants for tungsten and molybdenum. Values of 10^{-2} C are sketched as dotted lines on Fig. 7.4 corresponding to the four cases cited.

TABLE 1

Shear Modulus of Several Dislocations in W and Mo†

Material	Dislocation	Code	C, dynes-cm^2
Tungsten	Screw $a/2$ [111] ($\bar{1}\bar{1}0$)	W_1	10.01×10^{11}
	Screw a [010] ($\bar{1}01$)	W_2	16.04×10^{11}
Molybdenum	Screw $a/2$ [111] ($\bar{1}10$)	M_1	13.63×10^{11}
	Screw a [010] ($\bar{1}01$)	M_2	10.68×10^{11}

†After Reid.[10]

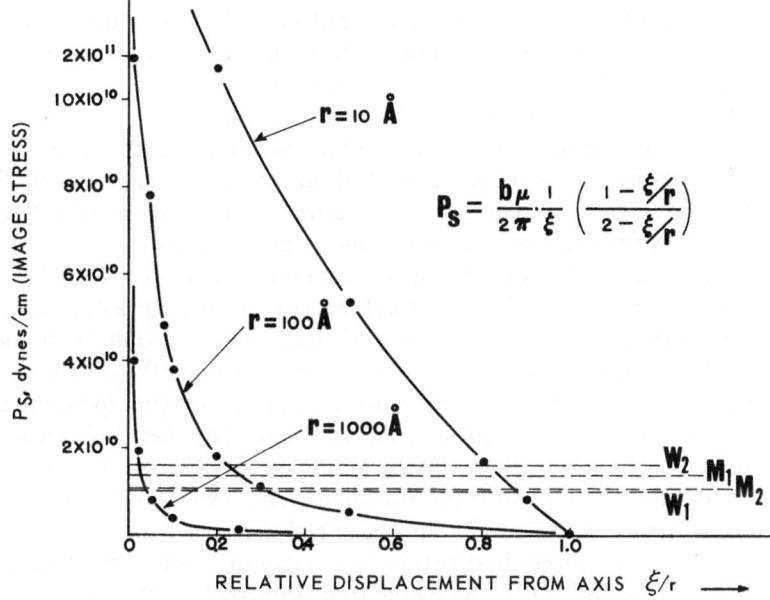

$$P_S = \frac{b\mu}{2\pi} \cdot \frac{1}{\xi} \left(\frac{1 - \xi/r}{2 - \xi/r} \right)$$

Fig. 7.4. Image force on pure screw dislocation parallel to axis of cylinder of radius r and displaced from it by ξ/r. Dotted lines correspond to 1/100 of shear modulus for tungsten (W_1 and W_2) and molybdenum (M_1 and M_2) in several crystallographic directions.

If we consider the specimen to be a series of thin concentric cylinders, as in Fig. 7.5, and examine a screw dislocation parallel to the common axis of these cylinders, we find that a displacement only slightly off the central axis of the topmost cylinder will tend to pull it farther. Segments of the dislocation deeper in the specimen will have progressively smaller image stresses

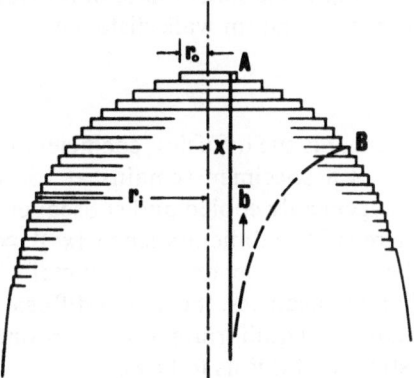

Fig. 7.5. Idealized cross section of field-ion specimen considered as a series of stacked thin cylinders. Screw dislocation initially at A would tend to assume position B as a consequence of image stresses.

until at some depth the image stress will not be sufficient to move the dislocation. The over-all tendency would be to go from a configuration like *A* to one like *B*. If the displacement *x* is sufficiently large, the dislocation will be swept out of the volume of the crystal actually imaged.

This analysis applies only to the pure screw and, once a configuration such as *B* is attained, we have a segment of mixed dislocation at an inclination to the specimen axis. There is no pressing incentive to solve the more general case at present, however, since one does not in any case know the Peierl's stress accurately and the local geometry is more complex than we have assumed. Nonetheless it is reasonable to expect the tendency to be the same in all cases since the total length of dislocation line would be reduced.

3. A still further complication arises if one considers the anisotropy of the elastic constants of nearly all real solids. Tungsten appears to be the only exception. Clearly, even though a specimen were truly hemispherical, the field-induced force field would result in significant nonhydrostatic stress (i.e., shear stresses) in a real anisotropic specimen. The result once again would be to favor dislocation motion, no doubt out of the crystal surface.

4. Finally, even purely hydrostatic stress can interact with all but pure screw dislocations, and it is conceivable that some dislocations might climb out of the crystal owing to the stresses alone, although the probability of this occurring seems fairly remote.[11]

All of these factors tend to lower the dislocation density in the imaged portion of the crystal and thereby reduce the probability of even finding a line defect that may then be imaged and analyzed. In an extreme case, it may also lead to gross plastic deformation of the tip or fracture.

Point defects. The situation for point defects is somewhat less involved than for line defects. The principal effect of specimen size is to increase the probability of vacancy migration out of the crystal surface.

A reasonable estimate of the importance of such effects can be obtained from the random walk distance

$$\sqrt{x^2} = \sqrt{2Dt} \tag{7.5}$$

where D is the diffusion coefficient of the material and t is time.

If a specimen remains at room temperature for 10^4 sec before being cryogenically cooled and subsequently imaged and D is, say, 10^{-20} cm^2/sec, there is little difficulty since, $(x^2)^{\frac{1}{2}} \cong$ one interatomic spacing. However, for the case of a quenched pure metal where there is a considerable supersaturation of vacancies, the actual diffusion coefficient may be significantly higher than the equilibrium one at room temperature. We can make a rough estimate of this as follows.

Assume that the concentration of quenched-in vacancies at room temperature is that corresponding to the quenching temperature T_q

$$\frac{n_v}{N} \cong \exp \left| -\frac{E_f}{kT_q} \right| \cong 10^{-4} \tag{7.6}$$

where n_v/N is the fraction of vacant lattice sites and E_f is the formation energy of a vacancy.

The frequency of atomic jumps for an atom-vacancy exchange mechanism may be estimated as

$$f = Zve^{-E_m/kT}e^{-E_f/kT} \tag{7.7}$$

where Z is the number of equivalent sites, v is an attempt frequency, and E_m is the migration energy. The second exponential term may be approximated by equation (7.6). Taking $v = 10^{12}$ cps and $Z = 10$, we have

$$f \cong 10^9 e^{-E_m/kT} \tag{7.8}$$

Since the diffusion coefficient D is given to reasonable approximation by

$$D = \frac{a_0^2 f}{6} \tag{7.9}$$

where a_0 is the interatomic spacing, we can estimate D for reasonable values of E_m. Thus for $E_m = 1.0$ eV, $D \cong 10^{-18}$ cm^2/sec corresponding to a random walk distance of 1 to 10 interatomic spacings. This would probably not be serious since sufficient field evaporation to obtain the end form would remove the affected volume. On the other hand, for either a lower E_m (say for divacancies) or with a microscope bakeout (to attain better vacuum) the volume over which vacancies would diffuse to the surface could comprise practically the whole imaging volume. For example, with $E_m = 0.8$ eV and all other conditions as above, $(x^2)^{\frac{1}{2}} \cong 100$ to 1000 Å, a 300°C bakeout yields an even larger diffusion path.

A further difficulty arises from the probable lowering of E_m when the specimen is imaged. Hydrostatic stresses are no blessing in this case, although cryogenic temperatures are a great aid. A specimen should never be allowed to heat up once an imaging sequence has been initiated.

Obviously some precautions must be taken if one wishes to count concentrations of point defects as singlets or as small groupings. One route is clearly to keep the specimen at as low a temperature as possible until it is imaged, except when absolutely impossible. Another possibility is to examine materials in which the energy of motion is relatively high (e.g., W and Pt). Still a third is to conduct an experiment on the specimen in the microscope (although this introduces additional unknowns). In any event, studies of interstitials are, apart from interpretational uncertainties, more prone to be influenced by the factors just described because of their lower migration energies.

On the other hand, if the point defects have coalesced to form secondary defects, the situation is much better. Except for large voids, secondary defects are simply a special case of the situation encountered when imaging dislocations. Voids, or large clusters, are treated at length in Section 11.10 dealing with FIM studies of radiation effects, but we might say simply here that, if the specimen retains its integrity (which it apparently can, see Fig. 8.4), then the only real problem is correct image interpretation. Dislocation loops and stacking fault tetrahedra should also be quite stable in the specimen both before and during imaging. A considerable flux of vacancies would be required to permit sufficient climb to eliminate defects of this kind, and they are otherwise relatively immobile. It should be noted that rather large climb forces are possible; however so little has been published on the topic that only speculation is possible at present.

7.3   Favorable Conditions for Imaging Defects

There are many factors tending to lower the dislocation density in a FIM specimen. Under what circumstances then might we expect to find line defects?

As mentioned briefly above, dislocations that are nearly colinear with the axis of large specimens would be in a very favorable position to be retained in the specimen under imaging conditions. This is so because these dislocations: (1) would be least affected by dislocation image forces, (2) would be least affected by shear stresses resulting from any of the factors already described, (3) would not lower their total line energy significantly by moving to other positions in the specimen, and (4) would be least affected by climb processes since their line tension would tend to bring them back, on the average, to the center. Furthermore, the width of such dislocations should not be drastically altered by the hydrostatic component of the imaging stresses. A significant fraction (more than chance) of reported dislocation images fall into this class, which lends credence to this analysis. An example is given in Fig. 8.11.

If two or more dislocations interact to form a node, the probability of obtaining an image of them should be considerably enhanced. The more complex the node, the better will be the chance of retaining all of its components, since these are effective pinning points. Stacking fault tetrahedra, e.g., offer a splendid chance of being stable under imaging conditions; however, here one need worry somewhat about the probability of encounter in the imaged volume. Too large a tetrahedra, e.g., would not be contained in a typical specimen if one could even find it. Mention of the phrase *effective pinning points* when referring to dislocation nodes, at once suggests a number of ways to stabilize line defects for observation in the FIM.

It is of course possible to impede dislocation motion in many ways, and all of these would be helpful to a greater or lesser degree in preventing dislocation escape at the surface. For example, grain boundaries are formidable barriers to dislocation motion, and one might expect to find line defect images at or near them. Indeed observations of spirals at grain boundaries have been interpreted in just this way (see Chapter 9). Substitutional solute atoms are also known to be effective dislocation "pinners," but they also frequently result in dissociation and these are discussed in Chapter 10. Interstitial atoms also lock dislocations strongly. As yet no significant use has been made of this property in FIM studies. Nonetheless, interstitial pinners offer a possible means of preserving the bulk dislocation density in an FIM specimen under imaging conditions. Other possible dislocation locking agents or barriers to motion include: (1) point defects produced by quenching or irradiation, (2) point defect agglomerates produced by irradiation, (3) finely dispersed second-phase particles, (4) short-range order, and (5) long-range order.

It should be mentioned that the problem of line defect rearrangement and loss in thin foils used in transmission electron microscopy has been quite successfully dealt with by neutron irradiation of deformed bulk crystals followed by thinning.[12] Significant differences in the appearance of the micrographs of otherwise identically treated specimens have resulted from using or not using the radiation-induced point defects as pinners.

7.4   Image Interpretation

Having described in some detail the many assorted difficulties associated with preparing a specimen representative of the problem to be studied, we are faced finally with the obstacle of image interpretation itself. We have already touched briefly on this in Section 7.2.1; however, let us examine what has so far come from experiments, and consider how it can be evaluated.

7.4.1   Point Defects

The difficulties associated with making meaningful counts of single and divacancies and small groupings are treated extensively in Chapter 11. We shall not treat this again here. Substitutional impurities are also covered in Chapter 10; however some comments about interstitial atoms are in order here.

There is still considerable uncertainty about the probable appearance and stability of an interstitial atom (self or impurity). It has generally been accepted that an interstitial will not be stable on the surface if it is uncovered by field evaporation since it would have a low coordination number. It is

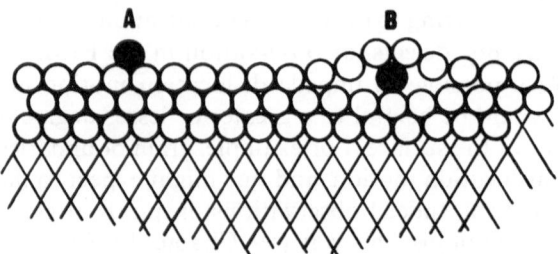

Fig. 7.6. Two possible imaging states for an impurity atom (dark) illustrating the bulge hypothesis and the exposed state.

thus assumed that an indication of an interstitial's presence would occur during the penultimate field evaporation step as a result of a bulging of the lattice atoms just above it at the surface. Figure 7.6 schematically shows two typical imaging states. In this figure atom *A* would presumably be unstable, while atom *B* would be visible indirectly in the image because of the bulge at the surface. There seems to be reasonable grounds for accepting this interpretation at present, but there are also other plausible explanations. Since it is possible for some adsorbed atoms to remain metastable on the surface and since many artifacts are associated with metastable surface sites (see Section 6.1), it is not unreasonable to expect at least some kinds of interstitials to remain metastably attached to the surface under the imaging conditions as well. In fact, interstitials have small atomic radii compared with the host atoms and would be somewhat screened from the field desorption stresses if they remained in interstitial sites. The appearance of large bright spots in the image can just as easily be associated with interstitials directly bound to the surface, as in the bulge hypothesis, providing a region of higher local ionization probability exists. In short, this is a difficult interpretational point to resolve at present. The answer will undoubtedly come from controlled and extensive studies of a variety of impurities with the use of statistical sampling methods. Such work is currently in progress at several laboratories.

7.4.2   Line Defects

As observed by Müller in his classic review of 1960,[1] "... the intersection of a single dislocation line with the surface is not very spectacular, and one has to scan the image carefully to find the imperfection." We have also seen (Section 7.2.1) that the atomic geometry at the intersection of a dislocation with an arbitrary surface is difficult to visualize. Coupled with this is the likelihood of image distortion near the defect (in the sense of variable local magnification and perhaps blurring or streaking due to local perturbations in the interatomic potentials) and even local atomic rearrangements due to the imaging stress field. All of these factors have tended to minimize the number of useful quantitative analyses of dislocations that have appeared in the literature to date.

Müller's[13] analysis of paired screw dislocations down the axis of an iron whisker is perhaps the best-known example. Drechsler, Pankow, and Vanselow[14] reported very early on observations of spiral structures in tungsten, but their measurements were with hydrogen ions at 1000°C prior to the advent of low-temperature field-ion microscopy and their interpretations have been strongly criticized as relating to steps considerably larger than atomic dimension, most likely caused by low index planes stacked off-center.[15] This criticism is probably correct; however the latest evidence indicates that all dislocations intersecting an arbitrary surface can give rise to a spiral structure as part of a long-range effect. We differentiate long-range from short-range here by reserving the latter as applying to the vicinity of the dislocation core, whereas the former refers to displacements normal to the imaging surface and results in spirals of a predictable type extending over 10 to 100 interatomic spacings.

Cottrell[16] points out the importance of the unending helicoidal surface resulting from the intersection of a dislocation and a free surface. He notes that this can result not only in the case of a pure screw dislocation, but for mixed and even edge dislocations provided the circumstances are right. Pashley[17] has recently applied this reasoning to field-ion images in a qualitative manner.

Ranganathan has carried the analysis further and has proposed that spirals resulting from dislocations emerging at a surface can be used to understand qualitatively the composition of low-angle grain boundaries.[2] His analysis has applicability in fact to all defect structures. He defines $\mathbf{N}$ as a vector describing the dislocation line, $\mathbf{b}$ as the Burgers vector of the dislocation, and $\mathbf{g}$ as the plane normal. In fact $\mathbf{g}$ is a reciprocal lattice vector which by definition has magnitude $1/d_{hkl}$. For the sake of clarity it may be written

$$\mathbf{g}_{hkl} = \frac{\mathbf{n}_{hkl}}{d_{hkl}} \tag{7.10}$$

where $\mathbf{n}_{hkl}$ is a unit normal vector to (hkl) and d_{hkl} is the interplanar spacing. A note of caution is in order here. This is a purely geometrical definition, and scattering theory should not be assumed to be applicable simply because the reciprocal lattice vector $\mathbf{g}_{hkl}$ has been used. One need only remember that the reciprocal lattice may be constructed by a pure geometric manipulation based on the primitive cell.

Ranganathan considers two cases:

$$\mathbf{N} \cdot \mathbf{g} \neq 0 \qquad \mathbf{g} \cdot \mathbf{b} = 0 \tag{7.11a}$$

and

$$N \cdot g \neq 0 \qquad g \cdot b \neq 0 \tag{7.11b}$$

The condition $N \cdot g \neq 0$ (both cases) corresponds to the dislocation line running at some angle to the plane, i.e., it does not lie in the plane. The criterion $g \cdot b = 0$ corresponds to the Burger's vector lying in the plane (hkl) and the displacement's being therefore confined to it. This configuration would be nearly invisible, as we discussed earlier, since the net planes would close on themselves and the displacements would be localized to a few atomic distances.

If $g \cdot b \neq 0$, however, the plane edges (rings) will not close on themselves and a spiral configuration will develop. The spiral depends *only* on the vertical component of the Burger's vector (the component normal to the plane intersected) and is always an integral multiple of interplanar spacings. This must be so since the Burgers vector (of an undissociated dislocation) by definition connects two lattice points. The integer multiple of d spacings will determine the number of leaves of the spiral, and the sense of the Burgers vector will determine the sense of the spiral (clockwise or counterclockwise).

This concept may be made quantitative as follows:

By using equation (7.10) and assuming $b = a_0[uvw]$

$$g_{hkl} \cdot b = \frac{n^{hkl}}{d_{hkl}} \cdot b = [hu + kv + lw] \tag{7.11}$$

The quantity in brackets is always a positive or negative integer for whole dislocations; therefore, the dot product $g \cdot b$ may be interpreted as the projection of b_{uvw} on n_{hkl} in units of d_{hkl}. This is Illustrated in Fig. 7.7, where either a single or double spiral can be made to originate on the (220) plane depending on the Burgers vector chosen. Thus

$$g_{(220)} \cdot b_{(\bar{1}\bar{1}0)} = -2 \tag{7.12a}$$

and

$$g_{(220)} \cdot b_{(\bar{1}0\bar{1})} = -1 \tag{7.12b}$$

Reversing either b would only change the sign of the integer, i.e., the sense of the spiral.

Tables 2 and 3 are summaries of possible spirals produced on selected low-index planes by $b = a_0/2 \langle 111 \rangle$ and $b = a_0/2 \langle 110 \rangle$ dislocations in b.c.c. and f.c.c., respectively.

This analysis can be extended to predict the spiral configuration expected from any line defect, including partial dislocations. In the latter

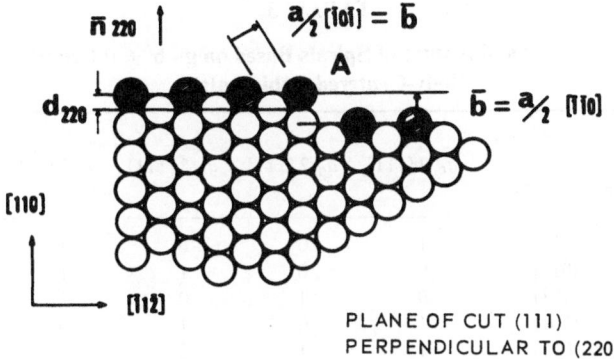

PLANE OF CUT (111)
PERPENDICULAR TO (220)

Fig. 7.7. Illustrating the origin of spiral structures when a dislocation intersects a plane. Either a single or double spiral may result, depending only on whether $\mathbf{b} = a_0/2\,[\bar{1}0\bar{1}]$ or $\mathbf{b} = a_0/2\,[\bar{1}\bar{1}0]$.

case, however, one obtains a fractional parameter for each partial dislocation. Furthermore a stacking fault is associated with the partial dislocations.

Computer simulation of these defect structures using an extension of the method developed by Moore (*cf.* Chapter 5) is underway in the author's

TABLE 2

Expectancy and Nature of Spirals Based on $\mathbf{g} \cdot \mathbf{b} \neq 0$ Criterion, Face-Centered Cubic Lattice

g \ b	$a_0/2\,[110]$	$a_0/2\,[1\bar{1}0]$	$a_0/2\,[101]$	$a_0/2\,[10\bar{1}]$	$a_0/2\,[011]$	$a_0/2\,[01\bar{1}]$
(220)	2	0	1	1	1	1
(02$\bar{2}$)	1	1	1	1	0	2
(20$\bar{2}$)	1	1	0	2	1	1
(202)	1	1	2	0	1	1
(022)	1	1	1	1	2	0
(111)	1	0	1	0	1	0
($\bar{1}$11)	0	1	1	1	1	0
(1$\bar{1}$1)	0	1	1	0	0	1
(11$\bar{1}$)	1	0	0	1	0	1
(200)	1	1	1	1	0	0
(020)	1	1	0	0	1	1
(002)	0	0	1	1	1	1

NOTE: 0 = no spiral, 1 = single leaved, 2 = double leaved.

TABLE 3

**Expectancy and Nature of Spirals Based on g · b ≠ 0 Criterion,
Body-Centered Cubic Lattice**

g \ b	$a_0/2$ [111]	$a_0/2$ [$\bar{1}$11]	$a_0/2$ [1$\bar{1}$1]	$a_0/2$ [11$\bar{1}$]
(110)	1	0	0	1
(01$\bar{1}$)	0	0	1	1
(10$\bar{1}$)	0	1	0	1
(101)	1	0	1	0
(011)	1	1	0	0
(1$\bar{2}$1)	0	1	2	1
(11$\bar{2}$)	0	1	1	2
(21$\bar{1}$)	1	1	0	2
(211)	2	0	1	1
(112)	2	1	1	0
(121)	2	1	0	1
(020)	1	1	1	1
(200)	1	1	1	1
(002)	1	1	1	1

NOTE: 0 = no spiral, 1 = single leaved, 2 = double leaved.

laboratory at present. There is every reason to expect that such studies, coupled with careful examination of experimental images, will go a long way toward making the analysis of dislocation images quantitative. Some recent results of such computer simulation are illustrated in Fig. 7.8a to d. The example illustrated here is idealized since the dislocation is a pure screw and intersects the center of a {420} in every case (a highly improbable situation experimentally). Nonetheless the confirmation of the expected spiral configuration is strikingly demonstrated. Such an approach may possibly even shed further light on the core structure of dislocations; however a better understanding of the forces present under imaging conditions is also essential.

7.5   Summary

Although there are considerable obstacles to an atomic level analysis of point and line dislocation structures with the FIM, considerable progress has been and is being made. A detailed understanding of the imaging criteria and of the forces acting prior to and during imaging is essential to ultimate utilization of the technique. Many useful studies may already be conducted, however, and the scope of these is bound to widen as the physics of image formation is better understood.

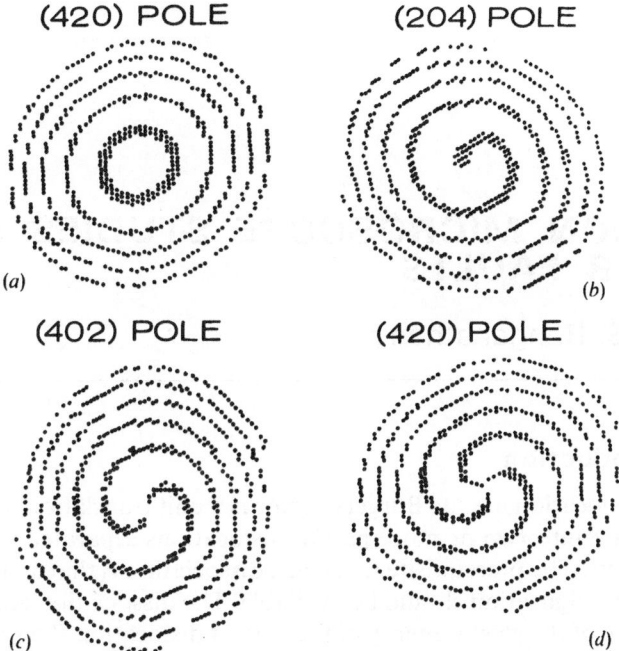

Fig. 7.8. Computer simulation of the long-range spiral structure resulting from the intersection of a pure screw dislocation ($\mathbf{b} = a_0/2\,[220]$) with the centers of $\{420\}$ planes in f.c.c. (a) Perfect crystal, (b) $\mathbf{g} \cdot \mathbf{b} = 1$, (c) $\mathbf{g} \cdot \mathbf{b} = 2$, (d) $\mathbf{g} \cdot \mathbf{b} = 3$. (After Sanwald, Ranganathan, and Hren.[3])

References

1. E. W. Müller, *Advan. Electron. Electron Phys.* **13**: 83–179 (1960).
2. S. Ranganathan, *J. Appl. Phys.* **37**: 4346 (1966).
3. R. C. Sanwald, S. Ranganathan, and J. J. Hren, *Appl. Phys. Lets.* **9**: 393 (1966).
4. W. T. Read, *Dislocations in Crystals*, McGraw-Hill (New York), 1953, pp. 115–116.
5. M. Drechsler and H. Liepack, *Colloq. Intern. Centre Natl. Sci (Paris)* **152** (1965).
6. E. W. Müller, *5th Field-Emission Symposium*, Chicago, 1958.
7. M. Drechsler and P. Wolf, *Proc. Intern. Congr. Electron Microscopy, 4th*, Springer-Verlag (Berlin), 1958.
8. E. W. Müller, *Direct Observation of Imperfections in Crystals*, Wiley (Interscience), 1962, pp. 77–99.
9. R. Bullough, "Dislocations," *AERE PGEC/L* 33 (Sept. 1964).
10. C. N. Reid, *Acta Met.* **14**: 13 (1966).
11. K. M. Bowkett, Ph.D. Dissertation, University of Cambridge, May, 1966.
12. U. Essmann, *Phys. Stat. Sol.* **3**: 932 (1963).
13. E. W. Müller, *J. Appl. Phys.* **30**: 1843 (1959).
14. M. Drechsler, G. Pandow, and R. Vanselow, *Z. Physik. Chem. (Leipzig)* **4**: 17 (1955).
15. E. W. Müller, *Acta Met.* **6**: 620 (1958).
16. A. H. Cottrell, *Theoretical Sctructural Metallurgy*, E. Arnold & Co. (London), 1955, pp. 231–232.
17. D. W. Pashley, *Rept. Progr. Phys.* **28**: 291 (1965).

Chapter 8

FIELD-ION MICROSCOPE STUDIES OF PLANAR FAULTS

S. Ranganathan†

8.1 Introduction

A perfect dislocation, i.e., its Burgers vector is a unit translation in the lattice, can dissociate into two or more partial dislocations separated by widths of stacking faults. The prerequisite for such a dissociation is that a mechanically stable new configuration should be available. The dissociation reaction leads to a lowering of the elastic energy of the perfect dislocation. This is balanced, however, by the increase in misfit energy in the plane. Several books on dislocations give an excellent treatment of these ideas.[1]

Stacking faults can arise in a material through deformation, aggregation of vacancies (e.g., during quenching), and interstitials (e.g., during irradiation). They affect the x-ray line profiles and lead to broadening as well as peak shifts. However their contribution is not easily isolated from other factors. Thus x-ray line broadening studies[2] have been of limited value. Transmission-electron microscope studies[3] have been conspicuously more successful. The faults produce characteristic interference fringes and have been studied extensively in investigations on quenching. The image width of a dislocation is of the order of about 100 Å, and hence, if the fault width is lower than this, the evidence becomes ambiguous. Also the evidence for stacking faults in b.c.c. metals where the fault energy is expected to be high has been particularly scanty. Therefore field-ion microscopy has an important role to play in the observation of stacking faults of small widths and high energies.

The energy of the stacking fault is an important parameter. It affects the mode of deformation, e.g., high-energy stacking-fault materials permit easy cross slip. It can also affect stress corrosion. Methods of varying

† Postdoctoral Research Associate, Inorganic Materials Research Division, Lawrence Radiation Laboratory, University of California, Berkeley, Calif.

reliability exist for the evaluation of the energy. If field-ion microscopic observations can shed light on this aspect, it should be welcome.

In this article we shall be solely concerned with the geometry of the stacking fault and its probable contrast in field-ion micrographs. We shall begin with a consideration of stacking-fault geometry in f.c.c. and b.c.c. structures. At one time streaks observed in field-ion micrographs were interpreted as stacking faults.[4] The question of streak contrast[5] receives full discussion here. Ryan and Suiter[6] have interpreted a crossover effect in tungsten micrographs as stacking faults on {111} planes. This contrast effect can also be interpreted as arising from nodes in a dislocation network.[7] The last two sections deal with two structurally simple instances—stacking faults in the hexagonal close-packed (h.c.p.) structure and domain boundaries in ordered alloys.

8.2 Stacking Faults in the Face-Centered Cubic Lattice

The f.c.c. structure can be considered made up of the stacking sequence $ABCABC$ based on (111) planes. There are three types of partial dislocations associated with faults in this stacking sequence: a Shockley partial and two Frank partials. In the case of the Shockley partial, the Burgers vector lies in the plane of the fault: for a Frank partial the Burgers vector is not parallel to the fault, and hence it is sessile.

Figure 8.1a illustrates a Shockley partial with Burgers vector $a_0/6\,[1\bar{2}1]$ on the (111) plane. The figure is for the dislocation in edge orientation. The

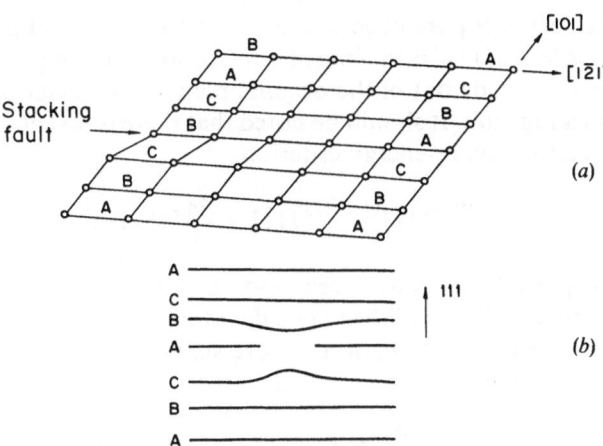

Fig. 8.1. (a) Illustration of a Shockley Partial with Burgers vector $-a_0/6\,[1\bar{2}1]$ (in edge orientation) with associated stacking fault in f.c.c. lattice. The plane of the figure is $(\bar{1}01)$ and the dislocation runs normal to this plane. Only one plane has been shown. (b) A Frank partial with associated stacking fault in f.c.c. lattice.

partial can arise from the following dissociation:

$$\frac{a_0}{2}[\bar{1}10] \rightarrow \frac{a_0}{6}[\bar{2}11] + \frac{a_0}{6}[\bar{1}2\bar{1}] \tag{8.1}$$

Figure 8.1*b* illustrates a Frank partial with Burgers vector $-a_0/3\ [111]$. Note that the partial is sessile. Also the fault is the same as in Fig. 8.1*a*.

Evidence for the existence of these partial dislocations and stacking faults rests on a firm experimental basis. Furthermore, interactions among the partials leading to the formation of Lomer–Cottrell locks and stacking-fault tetrahedra have been observed.[3]

8.3 Stacking Faults in the Body-Centered Cubic Lattice[1]

As detailed transmission-electron microscope studies of b.c.c. refractory metals have been lacking until recently, there is a great deal of confusion with regard to the presence of stacking faults in the b.c.c. lattice. Hirschhorn[8] has given an excellent summary of possible dislocation dissociations on {112}, {011}, and {310} planes.

The b.c.c. lattice can be considered built up of an ordered sequence of six planes, *ABCDEFAB*, based on {112}. Faults in this stacking sequence can arise from a dissociation of dislocations. Two reactions have received detailed attention.

$$\frac{a_0}{2}[111] \rightarrow \frac{a_0}{3}[112] + \frac{a_0}{6}[11\bar{1}] \tag{8.2}$$

The first partial is pure edge and sessile, and the second partial is pure screw and glissile. As $[11\bar{1}]$ is the line of intersection of three {112} planes, the second partial can glide out of the original plane of dissociation and create a dihedral stacking fault. It should be noted that there is no energy change on the Burgers vector square-energy criterion.

$$\frac{a_0}{2}[11\bar{1}] \rightarrow \frac{a_0}{6}[11\bar{1}] + \frac{a_0}{3}[11\bar{1}] \tag{8.3}$$

The two partials are pure screw and can once again glide out of the original slip plane. Sleeswyk[9] has considered an interesting variant of this reaction. If the original dislocation is pure screw, further dissociation into three partials can occur.

$$\frac{a_0}{2}[11\bar{1}] \rightarrow \frac{a_0}{6}[11\bar{1}] + \frac{a_0}{3}[11\bar{1}] + \frac{a_0}{6}[11\bar{1}] \tag{8.3a}$$

This results in a twofold symmetrical configuration of the three partials with one in the center and two at equidistant positions on different {112} planes.

Other types of dissociation on (112) planes have also been considered.

$$\frac{a_0}{2}[111] \rightarrow \frac{a_0}{6}[11\bar{1}] + \frac{a_0}{6}[133] + \frac{a_0}{6}[1\bar{1}1] \tag{8.4}$$

$$\frac{a_0}{2}[111] \rightarrow \frac{a_0}{6}[11\bar{1}] + \frac{a_0}{6}[113] + \frac{a_0}{6}[111] \tag{8.5}$$

Faults on other planes [e.g., (011) by Cohen et al.,[10] (310) by Segall[11]] have also been theoretically treated. Possible dissociations have been listed here in detail so that an attempt could be made to consider the contrast from stacking faults arising through these reactions from a theoretical point of view. There are few experimental observations of stacking faults in b.c.c. metals. Hirsch and Segall[12] found that the faults were on {310} planes in niobium. Nakayama et al.[13] have observed stacking faults in tungsten quenched from above 2000°C and derived a value of $\gamma_{sf} = 14.5$ ergs/cm^2 for impure tungsten. The work on alloys is slightly more definitive. Votava[14] has noted the existence of stacking faults on {112} planes in foils of molybdenum–35% rhenium.

8.4 Contrast from Partial Dislocations and Stacking Faults[1]

Moore[15] has tried to account for the pattern in a field-ion micrograph on the basis of pure geometry. A sphere was imagined to cut an infinitely extended crystal, and all the material outside the sphere was removed. A computer was used to plot in orthographic projection the positions of atoms that are within a certain distance from the surface of the sphere. The resulting pattern had a striking resemblance to the field-ion image. Moore presented a paper at the 11th Field-Emission Symposium in Cambridge, 1964, extending this approach to alloys and showed how the flexibility and the speed of the computer aided the choice of various possibilities leading to an image from the alloy. These results are summarized in the present volume in Chapter 5. This approach holds great promise and is capable of being extended to the surfaces of imperfect crystals. Some recent work along these lines is included in Section 7.4.2, but much remains to be done. Hence a qualitative approach will be adopted in the subsequent discussions.

When a perfect dislocation of Burgers vector **b** intersects the crystal surface on a particular net plane (**g** is the plane normal), a spiral will arise so long as $\mathbf{g} \cdot \mathbf{b} \neq 0$. This is similar to the visibility criterion in electron microscopy. The spiral is due to the component of the Burgers vector normal to the plane. If the component is equal to multiple interplanar spacing, a series of interleaved helicoids are expected. To follow Frank's terminology,[16] unlike the ideal crystal of n layers, the dislocated crystal consists of one layer only in the form of a helicoid. When the dislocation is of multiple strength, we have a crystal of several interleaved helicoidal layers. When it contains a number of dislocations, it consists of a number (possibly one) of similar interleaved *expanded Riemann surfaces*. Such information can be used for the determination of Burgers vector of a dislocation from field-ion micrographs.[17] (See also Section 7.4.)

The extension of these concepts to partial dislocations is not easy. A partial dislocation intersecting a close-packed plane will still be invisible if $\mathbf{g} \cdot \mathbf{b}_p = 0$ (for an exception in rhenium see Section 8.7). Thus an extended dislocation on the (111) plane can be missed on the net planes of (111). However, a stacking fault on any one of $(\bar{1}11)$, $(1\bar{1}1)$, or $(11\bar{1})$ will be visible on the (111) plane rings. In electron microscopy, if $\mathbf{g} \cdot \mathbf{b}_p = \pm\frac{1}{3}$, then also the partial dislocation is invisible or gives very weak contrast.[3] It will be interesting to apply this rule to study contrast in field-ion microscopy.

The foregoing considerations entirely ignore the real situation in a specimen being studied by field-ion microscopy. The dislocations are emerging at the surface. The surface is under an electrostatic stress. The specimen is finally subjected to the process of field evaporation. These three factors should be remembered in interpreting the image of a defect. The first two are dealt with in Chapter 7 and the last has been considered by Brandon[18] in some detail in Chapter 3. The surface section taken by field evaporation does not remain smooth in the case of a defective lattice. The variations in binding energy and lattice geometry at a lattice defect always result in some modification of the surface section. The problem is somewhat analogous to that experienced with chemical etching: the preferential removal of some atoms at the core of the defect can lead to field enhancement effects on neighboring atoms, so that a field evaporation *etch pit* is formed at the defect. The successful interpretation of the field-ion image thus depends very much on the correct estimate of the extent of cooperative evaporation.

8.5   Streak Contrast

Bright streaks are frequently observed in the field-ion images of specimens which have been subjected to certain treatments. There was considerable discussion of the possible causes for these image imperfections at the 11th Field-Emission Symposium in Cambridge. A subsequent paper by Ranganathan et al.[15] catalogued the different types of streaks and went some way toward interpreting their origins. The following summary draws heavily on this paper.

8.5.1   Tungsten–Rhenium Alloys

Ralph and Brandon[4] came across the first reproducible observations of image streaking in their study of the tungsten–rhenium system. Small streaks were found in deformed tungsten–rhenium alloys (Fig. 8.2) which

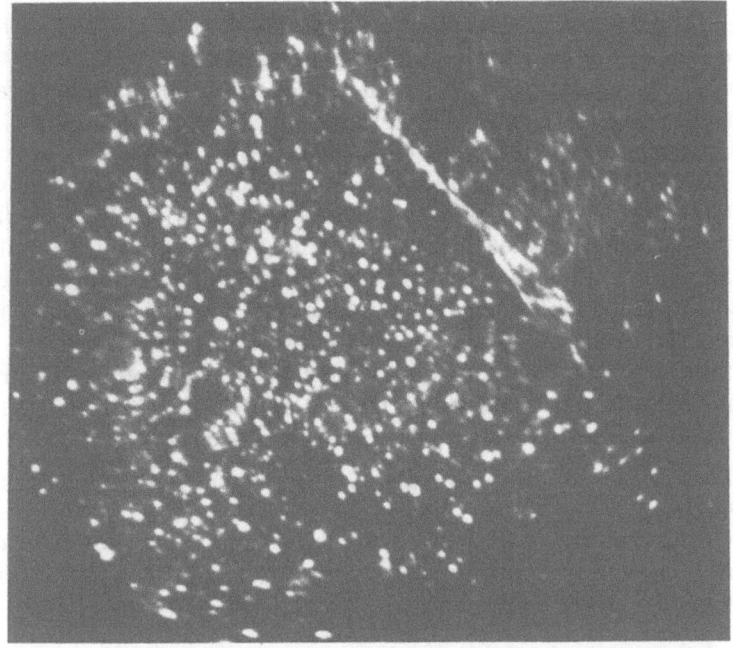

Fig. 8.2. Tungsten–5% rhenium showing bright lines (Ralph and Brandon[4]).

were interpreted as being due to the step produced where a stacking fault or a {112} plane intersects the crystal surface (Fig. 8.3). Preferential evaporation of atoms situated on the step was not thought likely since the atoms have the same coordination number and binding energy (though with a different disposition of neighbors) as other atoms giving rise to image points.

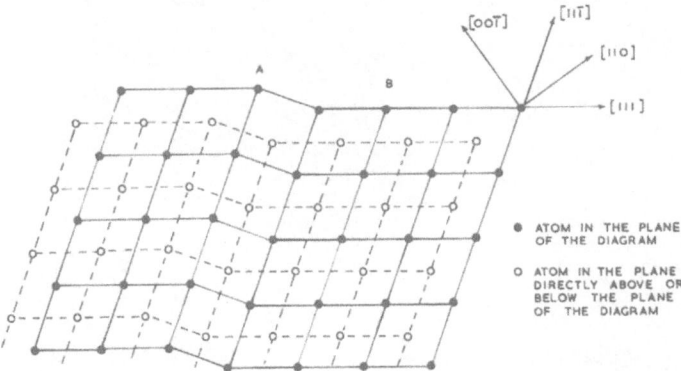

Fig. 8.3. A (1$\bar{1}$0) projection of a b.c.c. lattice containing a stacking fault and showing that a step is produced on the surface (Ralph and Brandon[4]).

Atoms on the step should however give bright images owing to field enhancement and focusing of field contours over the ridge formed by the fault.

Further consideration shows that a strong field enhancement factor
does not exist on (110) planes. The intersection of two faulted planes—the
diagram corresponds to the deformation fault with the stacking sequence
AB CD CD EFA—merely produces a kink on the (110) plane. While the
atoms at the kink may give slightly enhanced intensity, the effect will not
be the same as a step normal to the (110) plane. Such a step exists on planes
inclined to (110), e.g., (11$\bar{2}$). Also this deformation fault corresponds to the
first dissociation reaction described in Section 8.3 where a sessile $a_0/3$ [112]
dislocation forms the boundary for two fault planes. When through field
evaporation a number of atom layers have been taken off, most streaks
disappear. They do this in a gradual fashion and no special effect that will
indicate the existence of a sessile dislocation has been seen.

8.5.2   *Unambiguous Identification of Streaks: Image Superposition of Interfaces*

Image superposition can lead to streaking. Figure 8.4 shows streaking
arising from image superposition around the jagged edge of a surface crater
in tungsten.

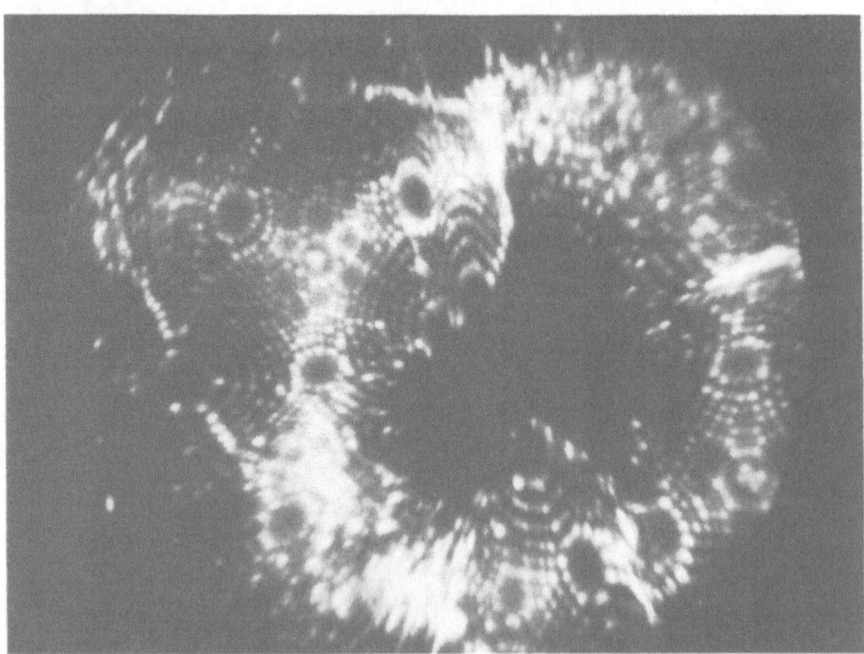

Fig. 8.4. Streaks arising from image superposition around the jagged edges of a surface crater
in tungsten (courtesy K. M. Bowkett) (from Ref. 5).

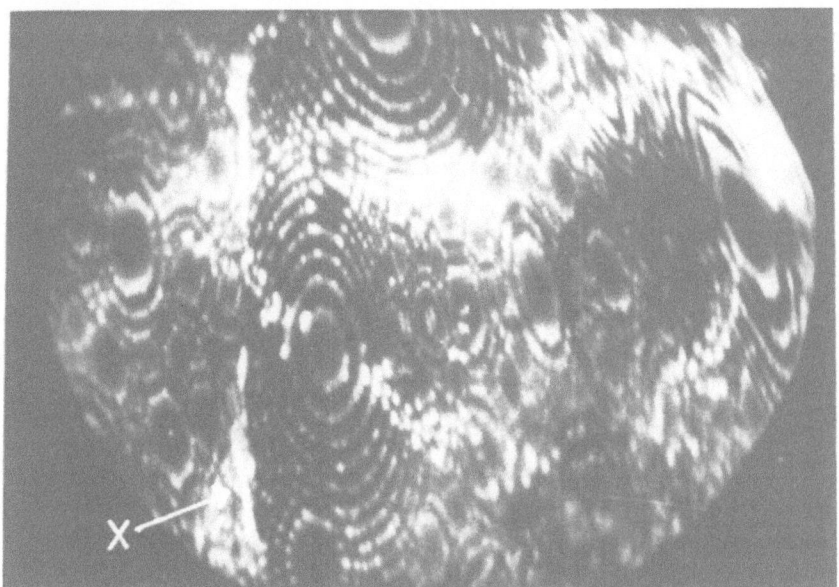

Fig. 8.5. Streak arising from a geometrical step on the surface at a grain boundary in tungsten (from Ref. 5).

Streaking at grain boundaries is fairly common. The presence of a grain boundary is readily detectable in the field-ion image because planes appear to be rotated out of their nomal positions on the image about the axis of misorientation. Thus there is little likelihood of streaks due to grain boundaries being confused with those generated by other mechanisms. Both electronic and geometrical factors can lead to such streaking. Figure 8.5 shows a streak arising from a geometrical step on the surface at a grain boundary in tungsten. Holland[19] has shown that localized electronic states exist in the vicinity of planar defects, including grain boundaries. Such electronic effects could well give rise to preferential ionization along the line of the boundary. Grain boundaries are also observed occasionally with a dark band (Brandon *et al.*[20]) lying along the boundary. These effects could well be connected.

8.5.3 Streaks in Asymmetric Tips: The Most Interesting Case

Asymmetric tips show some of the most drastic cases of streaking. The asymmetry of the tip is very striking. The radii from the central (110) to each of two neighboring $\{112\}$ poles often differ by a factor of 3. Sometimes the asymmetry is so pronounced that a measurement of the radii from the number of rings is not possible. Estimation from the appearance of high-index planes would lead to factors as high as 10.

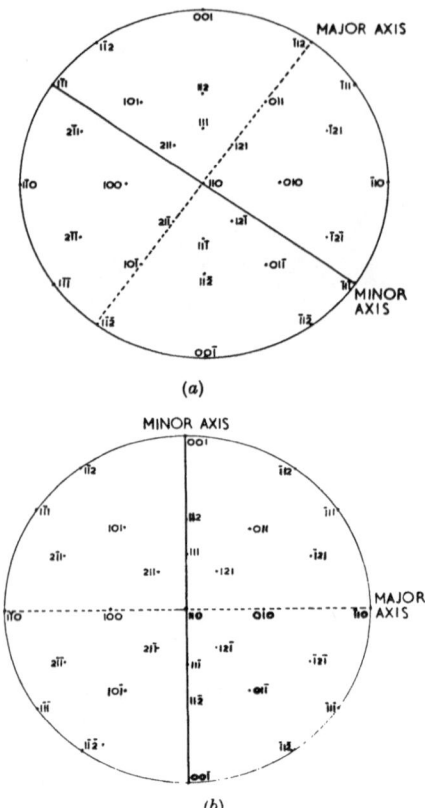

(a)

(b)

Fig. 8.6. Stereograms illustrating the crystallography of the two types of elliptical asymmetry found in tungsten (Ranganathan *et
al.*[5]).

There is a definite crystallographic nature to the asymmetry of the
specimens under consideration. The specimens have an elliptical cross
section and in general fall into two classes as illustrated in Fig. 8.6. Tungsten
has a preferred orientation which results in the tip axis invariably being
[110]. In one type of asymmetry the major and minor axes of the ellipse are
[$\bar{1}$12] and [1$\bar{1}$1], respectively. In the second less common type of asymmetry
the major and minor axes can be specified as [$\bar{1}$10] and [001], respectively.
Since in both cases the streaks lie perpendicular to the major axis of the
ellipse and also run parallel to the axis of the wire (as can be demonstrated
by field evaporation), the planes on which the streaks lie can be specified
accurately.

It is found that streaks lie on {112} or {110} planes, and it is significant
that these are the slip planes for the b.c.c. system. Examples of streaks lying
on {112} and {110} planes are shown in Figs. 8.7 and 8.8, respectively.

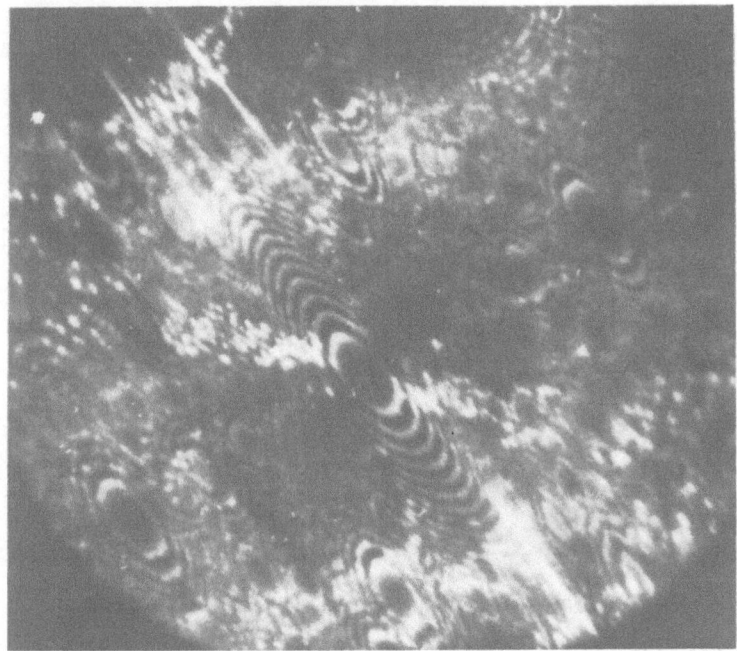

Fig. 8.7. Streaks lying on {112} planes in tungsten (from Ref. 5).

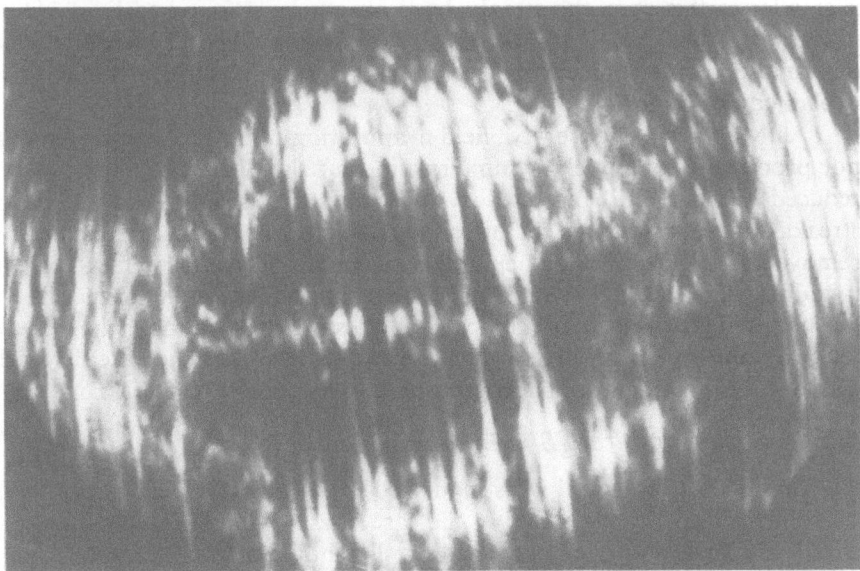

Fig. 8.8. Cold-worked neutron-irradiated tungsten. The streaks lie on {110} planes. (Courtesy K. M. Bowkett) (from Ref. 5).

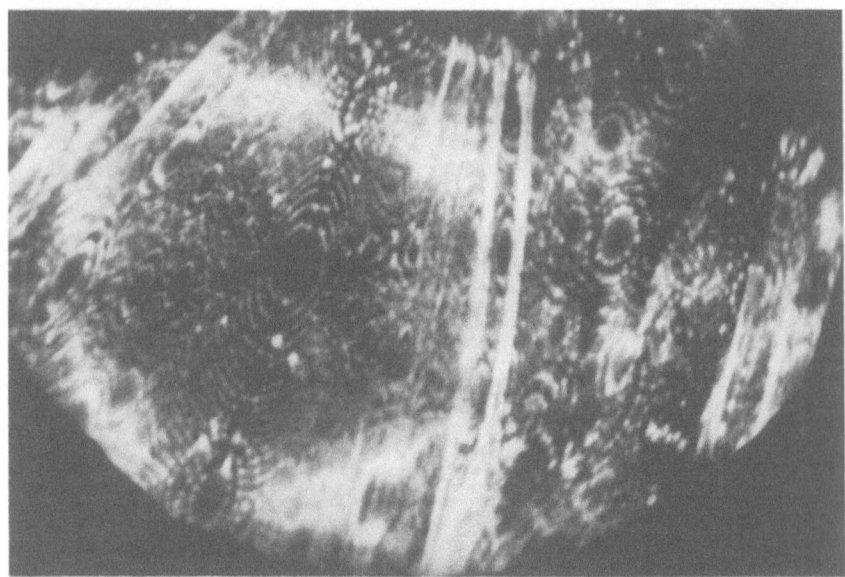

Fig. 8.9. Asymmetric iridium specimen. Streaks lie on {110} planes.

In the case of f.c.c. metals (e.g., platinum and iridium) asymmetry and associated streaking have also been observed. Although the body of evidence is smaller in this case, the streaks have always been found to lie on {110} planes (Fig. 8.9). The {110} planes in f.c.c. metals play no special part in deformation and the occurrence of streaks on these planes remains to be explained satisfactorily.

As Müller[21] has observed, field evaporation of an asymmetric tip does not produce a completely symmetric tip, but the degree of asymmetry is reduced. Similarly, when a tip containing streaks is field evaporated, the streaks and the associated distortion persist, although a few streaks may be removed. As a general rule, the number of streaks decreases with an increase in radius, which strongly suggests that the disturbance is confined to the top few hundred atom layers of the tip. An interesting observation is that streaks tend to split into two halves when the radius has been increased by field evaporation. Examples of such splitting are visible in Fig. 8.7.

8.5.4   Distinction from Slip Bands

Slip bands are easily distinguished from streaks. The shear stress component of the electric field in the case of an asymmetric tip can lead to slip. A corrugated surface produced by slip should assume a smooth shape after field evaporation, and this is at variance with the persistence of streaks. The crystallography of streaks in f.c.c. metals cannot be explained as well.

8.5.5 Contrast Theory

Basically streaks appear to be a planar type of imperfection. They are a gross structural feature in field-ion micrographs and are seldom clearly resolved. The most likely explanation is that streaks are a structural feature introduced during the process of specimen preparation. While both the electropolishing process and the field necessary for imaging can lead to the deformation of the specimen, the former is more important. With the use of a two-layer electrolyte, the detachment of the bottom portion of the wire at the end of the polishing process could lead to severe stresses and deformation of the metal. A rough estimate shows that the stress can reach values as high as 10^{10} dynes-cm^2. The asymmetry of the wedge-shaped specimen is again a function of the polishing process. As conditions of polishing vary from specimen to specimen, the random observation of streaks with no correlation with the state of the metal is not surprising. The effect of the imaging field on an asymmetric specimen is two-fold: The shear components of the field stress may lead to the deformation of the specimen, and the field variations over the elliptical surface can lead to a variable magnification in the image.

8.6 The Crossover Structure in Tungsten

A planar-defect structure[6] which has received considerable attention and can be interpreted with some confidence is what is called the *crossover structure* in tungsten (Fig. 8.10). It is easily recognized and well defined. There is no extra half-plane corresponding to the (011) planes. These net planes are seen to be drawn inward. Similar micrographs have appeared in earlier work on tungsten by Ranganathan[22] and on platinum by Müller.[23] Ryan and Suiter[6] have added the further remarkable observation that the effect disappeared over a hundred layers and then reappeared again in the same place. Their observations are described below.

From many observations it has been found that these defects show the following characteristics: Their maximum extent in the [2Ī1] direction is generally less than 30 Å, the lattice disturbance is usually evident for distances in the [011] direction of the order of 100 Å, and the spacing in the [011] direction between disturbed regions is some 200 to 400 Å. Sometimes the defect has been observed on a grain boundary (Fig. 8.11). These defects have been found in drawn wires as well as wires that have been heated to 700 to 1300°C while under stresses ranging up to fracture stress. However, the number of observations is not sufficiently large to enable their origin to be associated with any particular conditions of stress or temperature.

Ryan and Suiter interpret the crossover structure in terms of the diagram in Fig. 8.12a. They believe that it arises from an extended dislocation on the (111) plane. The dislocation reaction is thought to be similar to that proposed by Cohen *et al.*,[10]

$$\frac{a_0}{2}[\bar{1}1\bar{1}] \rightarrow \frac{a_0}{8}[\bar{1}0\bar{1}] + \frac{a_0}{4}[\bar{1}2\bar{1}] + \frac{a_0}{8}[\bar{1}0\bar{1}] \qquad (8.6)$$

If the dislocation extends by the movement of one of the $a_0/8\,[10\bar{1}]$ partials, faulting will occur on the $(11\bar{1})$ plane and the disturbance of the field-ion micrograph can arise from a number of effects due to the partial dislocations, the stacking fault, and the field-evaporation process. The three partial dislocations are all of mixed character so that, instead of having closed rings for the edges of the (011) planes in the vicinity of the extended dislocations, a type of spiral will occur, starting and finishing on the ends of the stacking fault and passing through each partial dislocation. The shear displacement across the stacking fault will produce a disturbance in the image if there is an appreciable component normal to the surface being examined. The extent of these effects may be exaggerated by the phenomenon of field evaporation. While the fault energy on a [111] plane should be high,

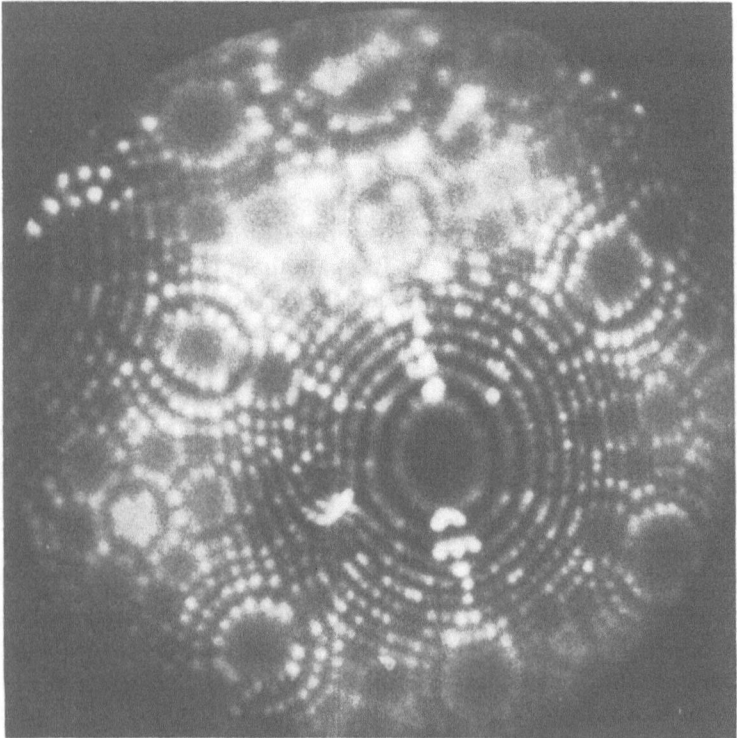

Fig. 8.10. Field-ion micrograph of characteristic defect structure in tungsten.

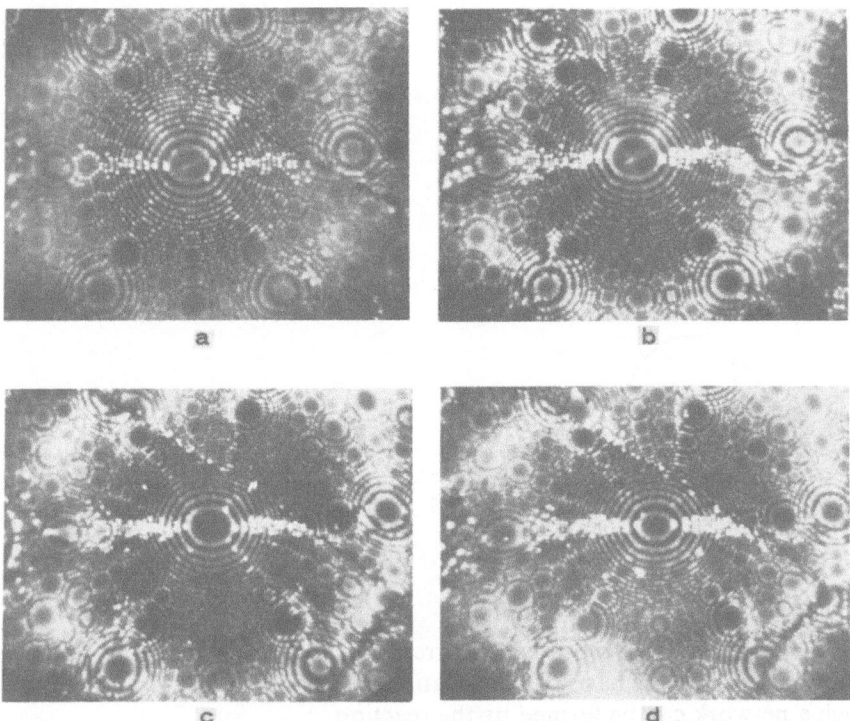

Fig. 8.11. The defect structure at a boundary in tungsten. The sequence brings out the appearance and disappearance of the structure. (Courtesy H. F. Ryan and J. Suiter.[6])

impurities in the specimen could have led to a segregation to the stacking fault and a consequent lowering in its energy.

The author[7] has made similar observations of crossover structures in tungsten (and iridium). The observational details generally confirm those of Ryan and Suiter. It was further observed that the effect could appear in a different place. Also in between such effects there was a residual contrast; in a particular case a single (011) plane was found to exhibit a closure failure.

In the model proposed by Ryan and Suiter, an edge dislocation lying on a {112} plane dissociates, and then the partials spread out on a {111} plane. This sequence appears improbable on energy considerations. Besides, there is no evidence for the presence of the partial dislocations with the given Burgers vectors. The reason for preferring the particular mode of dislocation dissociation appears to be that two of the partial dislocations could move on a {111} plane, the fault plane demanded by observation. The author has observed similar structures on both {111} and {100} planes. Hence a more general explanation appeared desirable.

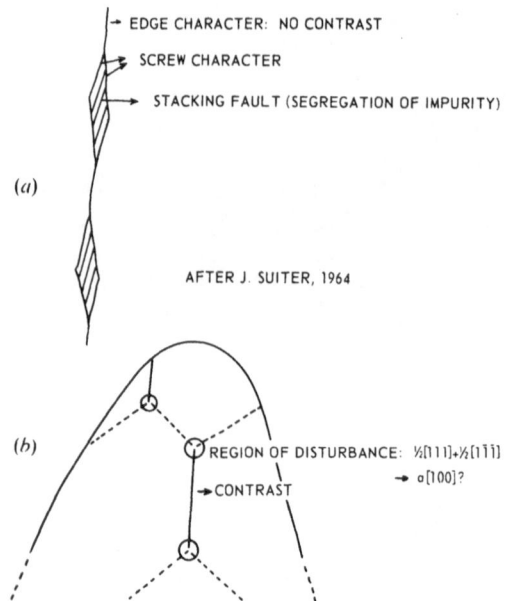

(a)

(b)

Fig. 8.12. (*a*) Schematic representation of proposed dislocation arrangement. (H. F. Ryan and J. Suiter[6]), and (*b*) proposed explanation on the basis of a dislocation network (S. Ranganathan[7]).

The regular and repeated occurrence of the defect seems to favor an interpretation on the basis of a dislocation network (Fig. 8.12*b*). For tungsten such a network can be formed by the reaction

$$\frac{a_0}{2}[1\bar{1}1] + \frac{a_0}{2}[11\bar{1}] \rightarrow a_0[100] \tag{8.7}$$

The locking-in of the dislocations might explain their stability in the presence of the field. In this interpretation, the disturbed structure would correspond to the asymmetrical threefold node formed by this reaction. Nets of this type observed in iron have been analyzed by Carrington *et al.*[24] In tungsten, also, such networks have been noted and a feature of many of these networks is the close spacing (about 100 Å) of the dislocations (F. O. Jones[25]).

8.7 Stacking Faults in Hexagonal Close-Packed Metals

The normal stacking sequence for h.c.p. metals is *ABAB* .... A fault involves three or more planes in the *ABC* order characteristic of f.c.c. All stacking faults in hexagonal crystals are on basal planes, so any given partial dislocation is confined to a basal plane.

Müller[26] has studied Re, Zr, Co, Zn, and Be. However only rhenium has received detailed attention. The basal plane has been observed by using single crystals grown in the *c* direction. Müller observed that net plane edges near the (0001) plane alternated in intensity due to the stacking sequence. Hence a stacking fault can be identified by its disturbance of this alternating effect. (The intensity alteration is spectacular and a complete explanation is not yet available.)

Müller[27] has also published hydrogen ion images of cobalt. Extensive studies of this metal should be of great interest as cobalt exhibits a phase transformation from h.c.p. to f.c.c. and has a low stacking-fault energy.[28]

8.8 Domain Boundaries

The field-ion micrographs from ordered alloys resemble those of metals in regularity and allow detailed interpretation.[29,30] In an ordered alloy antiphase domain boundaries are expected whenever two domains which are out of phase meet. The domain boundary is a kind of stacking fault in the superlattice and may end on a lattice dislocation of the disordered structure which becomes a partial dislocation of the superlattice. Southworth and Ralph[31] have presented examples of domain boundaries in cobalt–platinum. (See also Section 10.4.) The interface plane is (010), which provides the lowest energy condition. The boundary is identified as the antiphase structure that produces a characteristic bright spot/vacant site contrast at the boundary. If adjacent domains choose different cube axes of the parent grain for their (tetragonal) c axis, then a rotational domain boundary results when the domains meet. The paper by Southworth and Ralph[30] reports an example where the interface was lying on a $\{111\}$ plane.

Acknowledgments

The author would like to thank Mr. K. M. Bowkett, Dr. D. G. Brandon, Dr. B. Ralph, Dr. H. F. Ryan, and Dr. J. Suiter for the supply of illustrative material for this review. He is grateful to Professor G. Thomas for encouragement and helpful discussion and the United States Atomic Energy Commission for financial support.

References

1. A. H. Cottrell, *Dislocations and Plastic Flow in Crystals*, Oxford University Press (London), 1953; also W. T. Read, *Dislocations in Crystals*, McGraw-Hill (New York), 1953; and J. Friedel, *Dislocations*, Pergamon Press, (London), 1964.
2. B. E. Warren, *Progr. Metal. Phys.* **8**: 147 (1959).
3. G. Thomas, *Transmission Electron Microscopy of Metals*, Wiley (New York), 1962.
4. B. Ralph and D. G. Brandon, *Phil. Mag.* **8**: 919 (1963).
5. S. Ranganathan, K. M. Bowkett, J. Hren, and B. Ralph, *Phil. Mag.* **12**: 841 (1965).
6. H. F. Ryan and J. Suiter, *J. Less-Common Metals* **9**: 258 (1965).
7. S. Ranganathan, *J. Less-Common Metals* **10**: 368 (1966).
8. J. S. Hirschhorn, *J. Less-Common Metals* **5**: 493 (1963).
9. A. W. Sleeswyk, *Phil. Mag.* **8**: 1467 (1963).
10. J. B. Cohen, R. Hinton, K. Lay, and S. Sass, *Acta Met.* **10**: 894 (1962).
11. R. L. Segall, *Acta Met.* **9**: 975 (1961).

12. P. B. Hirsch and R. W. Segall, Substructure and Mechanical Properties of Crystals, *WADD TR*-61-181, 1961.
13. Y. Nakayama, S. Weissman, and T. Imura, in: *Imperfections in Crystals*, (Interscience) (New York), 1961, p. 573.
14. E. Votava, *Acta Met.* **10**: 745 (1962).
15. A. J. W. Moore, *Phys. Chem. Solids* **23**: 907 (1962).
16. F. C. Frank, *Advan. Phys.* **1**: 91–109 (1952).
17. S. Ranganathan, *J. Appl. Phys.* **37**: 4346 (1966).
18. D. G. Brandon, *AIME Conf. High-Temperature High-Resolution Microscopy, Chicago,* 1964.
19. B. W. Holland, *Phil. Mag.* **8**: 87 (1963).
20. D. G. Brandon, B. Ralph, S. Ranganathan, and M. Wald, *Acta Met.* **12**: 813 (1964).
21. E. W. Müller, *Acta Met.* **6**: 620 (1958).
22. S. Ranganathan, *Third European Regional Conference on Electron Microscopy, Prague,* 1964, p. 265.
23. E. W. Müller, *Proc. Intern. Conf. Electron Microscopy, 4th, Berlin* **1**: 820 Springer Verlag (Berlin), 1958.
24. W. Carrington, K. F. Hale, and D. McLean, *Proc. Roy. Soc. (London),* **A259**: 203 (1960).
25. F. O. Jones, *J. Less-Common Metals* **2**: 163 (1960).
26. E. W. Müller, *Third European Regional Conference on Electron Microscopy, Prague,* 1964, p. 161.
27. E. W. Müller, in: *Direct Observations of Imperfections in Crystals*, Wiley (Interscience) (New York) 1961, p. 77.
28. T. R. Anantharaman and J. W. Christian, *Acta Cryst.* **9**: 479 (1956).
29. E. W. Müller, *Bull. Am. Phys. Soc.* **7**: 27 (1962).
30. B. Ralph and D. G. Brandon, *Journées Int. des Applications du Cobalt*, June 9–11, 1964, p. 76.
31. H. N. Southworth and B. Ralph, *13th Field-Emission Symposium*, Cornell University, Ithaca, N.Y., 1966.

Chapter 9

FIELD-ION MICROSCOPE STUDIES OF INTERFACES

S. Ranganathan†

9.1 Introduction

Interfaces are of wide and varied occurrence. The boundary between two crystals forms just one type. The crystals themselves can differ in orientation, structure, and composition. All three kinds of interface crystals have been studied with the field-ion microscope. However extensive studies have been confined to boundaries separating crystals differing only in orientation. The primary problem in this field has been to evolve a model for the atomic configuration at high-angle grain boundaries. These attempts have met with some success. Other properties of grain boundaries are of interest and will be discussed; the scope for field-ion microscope studies seems to be limited however. The chapter concludes with observations of boundaries separating crystals differing in structure and composition.

9.2 Incidence of Grain Boundaries

In the transverse cross section ($\sim 10^{-4}$ cm^2) of a wire of 0.1 mm diameter, there are 10,000 grains for a grain size of 1 μ and thrice as many boundaries. The area imaged in the field-ion microscope is of the order of 10^{-10} cm^2. Hence the chances of imaging a grain boundary are expected to be low. In practice the chances are considerably improved as one samples a volume by the process of field evaporation. Also heating the tip *in situ* sometimes produces a migration of grain boundaries into the field of view. Simultaneous heating and deformation of the wire prior to preparation of the specimen, a suggestion by McLean, led to a small grain size and appreciably improved the chances of finding a grain boundary in the field of view. Certain re-crystallization treatments appear to be critical: tungsten wire heated to

† Postdoctoral Research Associate, Inorganic Materials Research Division, Lawrence Radiation Laboratory, Berkeley, Calif.

1000°C and molybdenum wire heated to 700°C were particularly profitable materials to work with.

Brock[1] carried out a pioneering study of grain boundaries with the field-emission microscope and followed a suggestion by Müller to increase the incidence of grain boundaries. The tungsten specimen was thinned, bent, and then a tip was sharpened at the bend. The tip was then annealed at 2400°K. It was believed that the prior deformation induces recrystallization and causes a number of grain boundaries to appear in the field of view. In a report Müller has claimed that this method led to the appearance of grain boundaries in 90% of the tips so treated. In an early study of grain boundaries with the field-ion microscope, Wolf[2] has made use of this technique. No extensive use of this method has been reported. A major drawback is that a tip will have a large radius after annealing and the loss of resolution at large radii is especially serious when liquid nitrogen is used for cooling.

9.3 Grain Boundary Analysis

An arbitrary grain boundary has five degrees of freedom. Three are needed to define the misorientation between the two crystals. These three degrees of freedom can be given as an angular misorientation around an axis. If it is assumed that the field-ion micrograph is a stereographic projection, standard crystallographic methods can be employed to determine the axis–angle pair. The first such determination with the field-ion micrograph was made by Ralph and Brandon.[3] A great circle is drawn through two poles with the same indices, one in each crystal. The mid-point of the arc between the two poles is linked to the pole of the great circle by another great circle. It is clear that the axis of misorientation lies on the second great circle. The intersection of this great circle with another similarly derived gives the axis of misorientation. Ralph and Brandon used any two poles. It seems preferable to operate with the cube poles arrived at after giving weight to the determination of the three poles (Fig. 9.1a). This graphical analysis has an accuracy within $\pm 2°$, and the assumption of a stereographic projection aggravates the inaccuracy. Fortunately the b.c.c. metals exhibit a strong wire texture; the misorientations are usually around [110], and this pole is in the center of the micrograph. Hence greater confidence can be attached to these determinations than would seem possible at first sight.

Two degrees of freedom are required to specify the orientation of the boundary plane. Once again simple geometrical relations can be used. (See, e.g., Hren.[4]) Figure 9.1b illustrates the situation. The inclination of the pole of the plane to the axis of the wire is equal to $90° - \tan^{-1}$ $(\gamma \sin \alpha)/[(N - n)h]$, where γ is the radius of the tip; α is the angle at which the pole (intersection of the boundary with the surface in a zone containing the axis of the wire) appears on one micrograph; N is the number of planes

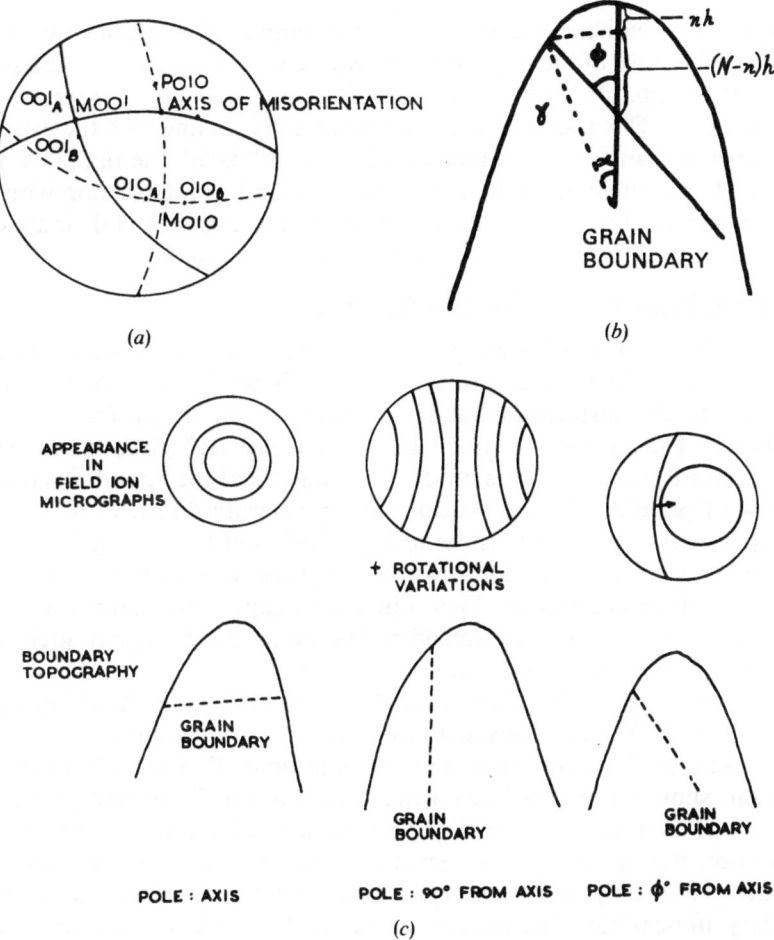

Fig. 9.1. (a) Determination of angle and axis of misorientation from the stereographic projection of a bicrystal, where *P* is a pole and *M* is the midpoint of arc joining corresponding poles; (b) Determination of the orientation of the plane of the boundary; (c) The three types of orientation of the boundary plane and their appearance on a stereographic projection.

of step height *h*, that are required to be removed so that the point of intersection shall reach the center of the micrograph; and *n* is the number of rings counted from the center to the pole (intersection). This expression is different from the one used by Hren.[4] One could distinguish three types of boundaries on the basis of this equation and the relation between N and n. (1) $N = \infty$, i.e., the intersection of the boundary with the surface does not seem to move with field evaporation. Most boundaries are of this type.

This appears to indicate that this is a stable configuration attained by grain boundaries because of the large kinematical freedom they enjoy in the small volume of the tip. (2) $N = n$. The pole of the boundary is the same as the axis of the wire. This is of unusual occurrence and is in line with the expectation that such transverse boundaries will tend to "flash" the tip. (3) $N \neq n$. The boundary is inclined to the axis of the wire, and its intersection with the surface is seen to move on field evaporation. All three cases and their appearance on a stereographic projection are illustrated in Fig. 9.1c.

9.4   Structure of Low-Angle Boundaries

The observation of etch pits at the entrance of dislocations and direct electron-microscopic observations of dislocation arrays have fully substantiated the dislocation model of Burgers and Bragg for low-angle boundaries. The review by Amelinckx and Dekeyser[5] gives an excellent and rigorous treatment of this model. A symmetrical tilt boundary consists of a row of parallel edge dislocations at appropriate distances to give the required misorientation. A twist boundary consists of a crossed grid of screw dislocations. More complicated arrays are required when there is a departure from these strict conditions. Field-ion microscope observations are of a confirmatory nature. They are considered below as they bring out interesting details about contrast in field-ion micrographs.

If there are to be five dislocations in the field of view, the dislocations will be about 60 Å apart. This distance of separation will correspond in the case of a simple tilt boundary to a misorientation of 3°. The grain boundary in iridium shown in Fig. 9.2 has a misorientation of 2° around [111]. The plane of the boundary was inclined to the axis of the wire, and on field evaporation the boundary intersected the surface at a different place. The asymmetry of the tip made an accurate determination of the plane of the boundary impossible. The streaks occur on $(01\bar{1})$ planes in f.c.c. metals. The intersection of the boundary is parallel to the streaks which shows that their poles occur in the same zone. This fact and the movement of the boundary intersection pointed to (120) as the boundary plane. Hence the boundary has twist character and must have screw dislocations making up the array.

The dislocation structure seen in Fig. 9.2 has the appearance of a classical screw dislocation. If the boundary plane is (111) and if the axis of misorientation is [111], the model for the boundary consists of a hexagonal grid of dislocations having Burgers vectors $a_0/2\,[1\bar{1}0]$, $a_0/2\,[\bar{1}01]$, and $a_0/2\,[01\bar{1}]$. When the orientation of the boundary plane is changed, e.g., to (120), there is only a change in the mesh shape and the pattern becomes such that its projection along the [111] on a (111) plane produces the same hexagonal net. The dislocation seen at the boundary must form part of such

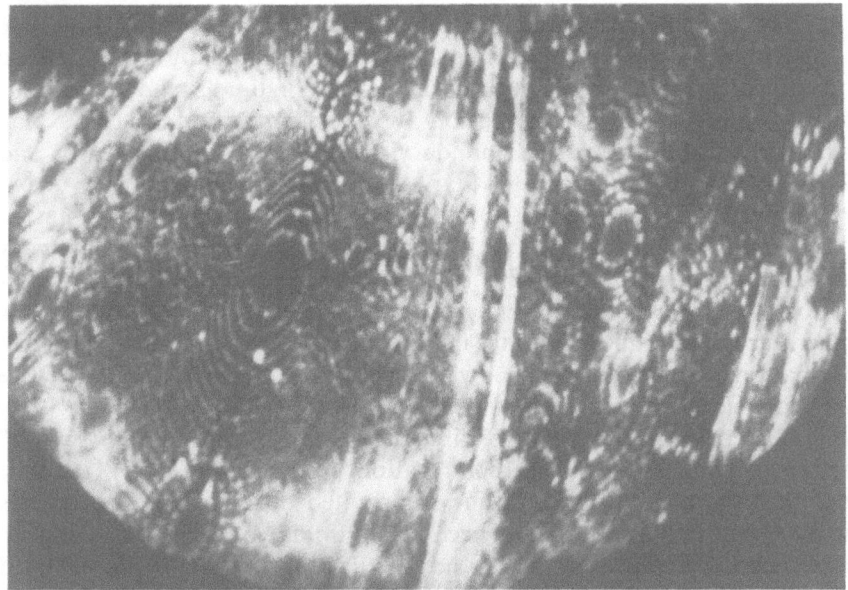

Fig. 9.2. A low-angle boundary with a misorientation of 2° around [111] in iridium.

a grid. However the Burgers vectors of the dislocation must be of the type $a_0/2\,[110]$ or $a_0/2\,[101]$ or $a_0/2\,[011]$, $a_0/2\,[1\bar{1}0]$, $a_0/2\,[\bar{1}01]$, and $a_0/2\,[01\bar{1}]$ all lie on the (111) plane and hence will not give rise to a spiral based on (111) planes. The dislocation is thus probably a stranger dislocation in the network.

Interlocked and interleaved spirals have been observed at low-angle boundaries in tungsten and molybdenum. They are not easily analyzed into the component dislocations.

9.5 Structure of Twin Boundaries

A twin boundary is said to be *coherent* when the twin plane and the composition plane coincide [(111) in f.c.c. and (112) in b.c.c.], and *noncoherent* when this condition is not fulfilled. In coherent twins the atoms in the composition plane are in the correct position in both lattices. The twin boundary energy is correspondingly low.

Müller[6] has presented an example of what appears to represent a small-angle twin boundary with (100) as the boundary plane. A mirror image is seen across this plane. The use of the phrase *small-angle twin* is not clear in this context. He has also observed a twin boundary in steel with (121) as the boundary plane.[7] This is presumably a coherent twin, and the atomic matching is excellent.

A few noncoherent annealing twins have been observed in tungsten. A particularly interesting example showing exceptionally good atomic matching has been analyzed in detail by Hren.[4] Figure 9.3 shows a similar twin in tungsten. The misorientation 180° rotation around [112] is seen very clearly. The plane of the boundary is a few degrees from $(\bar{1}13)_A$ and $(2\bar{2}3)_B$. The plane is of moderately high atomic density and will have one in three atoms in coincidence sites. Because of the deviations in the orientation of the boundary plane, several ledges were observed in the boundary (Fig. 9.3a). The structure of such ledges is discussed in a later section.

Twinning is a mode of deformation and is especially favored at low temperatures and high strain rates in b.c.c. metals (Hall[8]). In three instances deformation twinning of the tips was observed. Figures 9.4 and 9.5 were obtained before and after the twinning transformation. The poles [110] and [112] remain in the same position, as a twin can be related by a 70.5° rotation around [110] and a 180° rotation around [112]. It will also be noted that the zone decoration line has rotated through 70°. The tip was originally hemispherical and has become ellipsoidal as a result of twinning. There is considerable stress relief, and the tilt in the surface could have led to the observed streaking.

In another case twinning occurred as the voltage was raised to the value necessary for imaging. The boundary region appeared as a chasm at first and gradually this closed up. A cavity can still be seen in Fig. 9.6. The twin relationship is obvious. The spiral effect implies that a dislocation structure has been added to the boundary. The spiral is similar to the one observed at the low-angle boundary in iridium. The continuity of the spiral is interrupted at the ninth plane, where a second dislocation might be intersecting the surface. The twin has a noncoherent interface, and the plane of the boundary is very near $\{112\}_A$ and $\{110\}_B$. Both are close-packed planes in the b.c.c. lattice, and this explains the excellent atomic matching across the boundary. The dislocation observed near the central (110) pole will explain the slight deviation in the plane of the boundary from that in the close-packed planes. In this case the orientation relationship is exact, and hence there is no superposed small-angle boundary.

With field evaporation, the spiral remained in the same place, which showed that the dislocation was running in the (110) direction. Similar spirals arising through evaporation have been observed on a macroscopic scale by Votava and Berghezan.[9] They ascribed the spirals to twinning dislocations. It is difficult to see how a single twinning dislocation can give rise to a simple spiral. The suggestion of Ryan and Suiter[10] that these structures result from slip dislocation adsorbed at the boundary is attractive. However even in this case, because the crystals are rotated with respect to each other, the continuation of the atomic planes across the boundary is surprising.

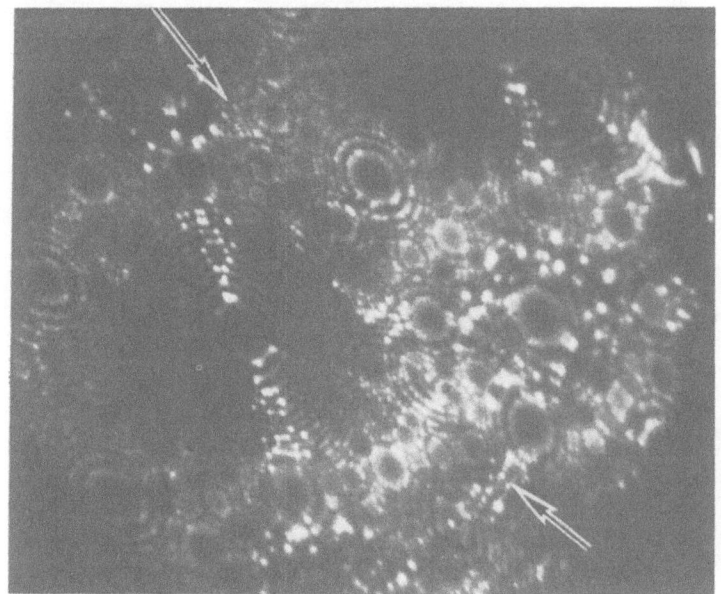

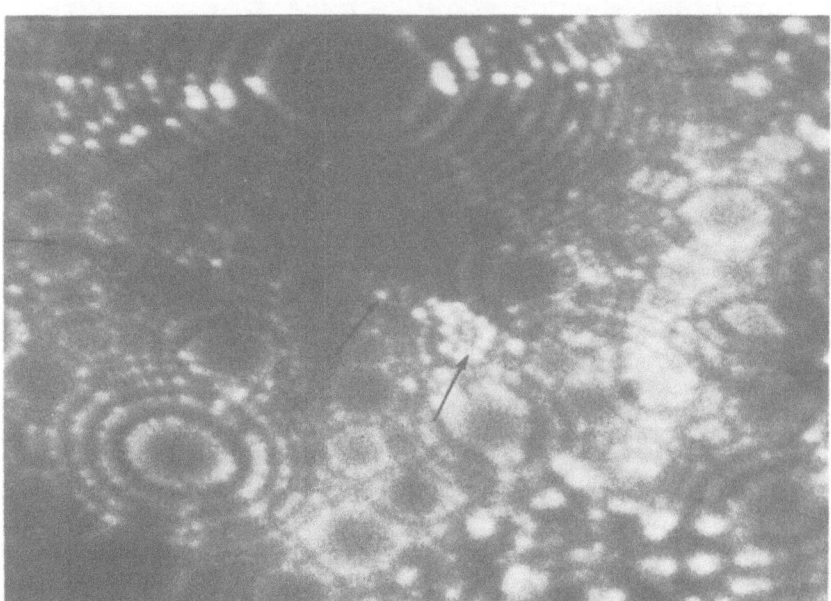

Fig. 9.3. A noncoherent twin boundary in tungsten.

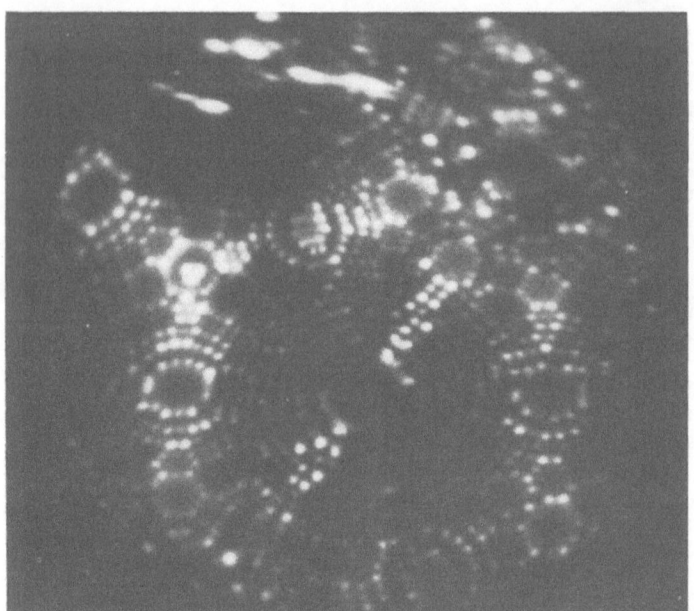

Fig. 9.4. Image from a tungsten tip before deformation twinning *in situ* occurred.

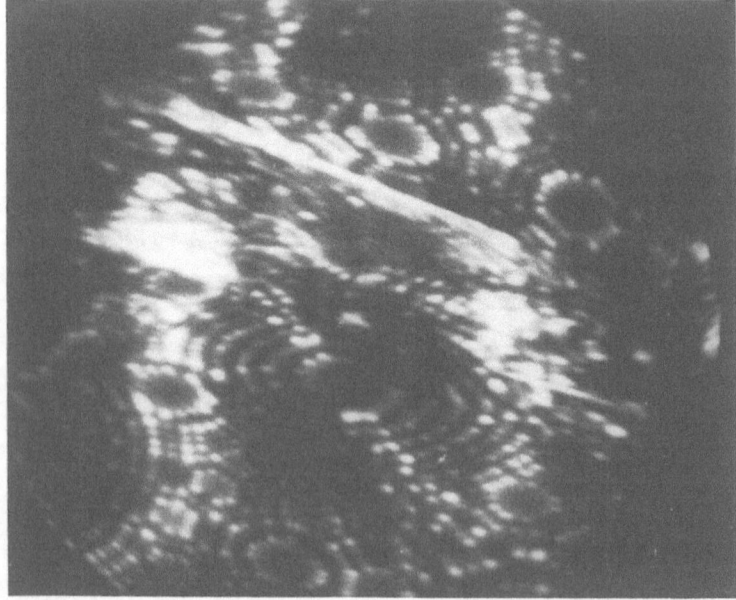

Fig. 9.5. Image from a tungsten tip after deformation twinning.

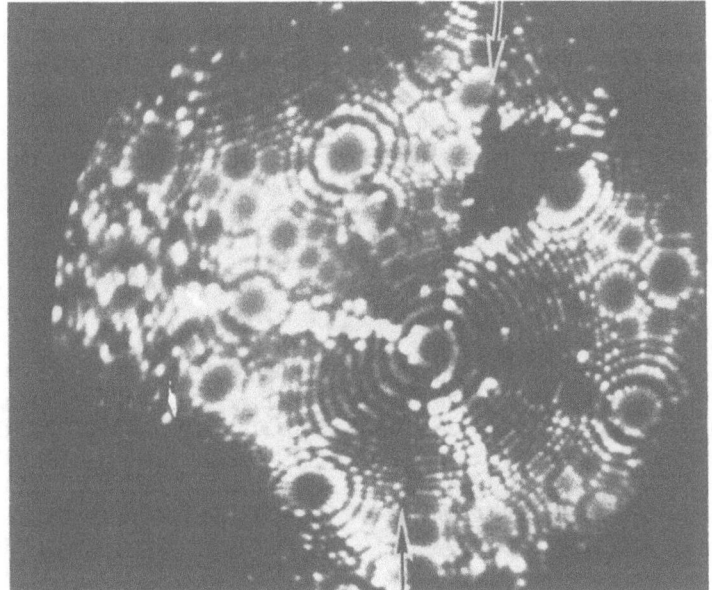

Fig. 9.6. A deformation twin interface in tungsten. Note especially good atomic matching.

9.6 Structure of Special and Random Boundaries

9.6.1 Theory

As a result of the field-ion microscopic observations of the structure of grain boundaries, a new model for the atomic configuration at high-angle boundaries has been evolved by Brandon, Ralph, Ranganathan, and Wald.[11] The relationships in the coincidence of lattice sites that form the basis of this theory were worked out by the author.[12]

Rotational symmetry operations on a lattice bring it into complete self-coincidence. However, partial coincidence can occur for certain other rotations about an axis. Two crystal lattices, related by such an angular rotation about an axis, have certain common sites that are located on a single lattice of larger cell dimensions. This larger lattice is called the *coincidence-site lattice*, and the boundary it gives rise to is termed a *special boundary*. Table 1 gives the axis-angle values which lead to coincidence-site lattices. Such special boundaries can minimize their energy by following planes of dense coincidence. (See Table 2.) A grain boundary running at a small angle to this plane will take up a stepped structure such that it has a maximum surface area in the close-packed planes of the coincidence-site lattice, which is analogous to a dislocation constrained to be along a

TABLE 1

Axis-Angle Pair and Corresponding Coincidence in the Cubic System

Axis of rotation A	Multiplicity Σ^*	Least angle or rotation ω, deg.	Axis of rotation A	Multiplicity Σ^*	Least angle of rotation ω, deg.
100	5	36.9	311	3	146.4
100	13_a	22.6	311	5	95.7
100	17_a	28.1	311	9	67.1
110	3	70.5	311	11	180.0
110	9	38.9	311	15	50.7
110	11	50.5	311	15	117.9
110	17_b	86.6	320	7	149.0
110	19_a	26.5	320	11	100.5
111	3	60.0	320	13_a	180.0
111	7	38.2	320	17_b	121.9
111	13_b	27.8	320	19_b	71.6
111	19_b	46.8	321	7	180.0
210	3	131.8	321	9	123.7
210	5	180.0	321	15	150.1
210	7	73.4	321	15	86.2
210	9	96.4	322	9	152.7
210	15	48.2	322	13_a	107.9
211	3	180.0	322	17_b	180.0
211	5	101.6	410	9	152.7
211	7	135.6	410	13_b	107.9
211	11	63.0	410	17_a	180.0
211	15	78.5	411	9	180.0
221	5	143.1	411	11	129.6
221	9	90.0	411	17_a	93.4
221	9	180.0	411	19_b	153.5
221	13_b	112.6	331	5	154.2
221	17_b	61.9	331	7	110.9
310	5	180.0	331	11	82.1
310	7	115.4	331	17_b	63.8
310	11	141.9	331	19_a	180.0
310	13_b	76.7			
310	19_a	93.0			

*Σ's 13, 17, and 19 each generates two systems, and these are distinguished by the subscripts a and b.

direction that is not a low-energy direction. This step structure is shown in Fig. 9.7 at *B–C*. This figure is a two-dimensional model for a b.c.c. bicrystal with a coincidence-site density of 1 : 11. The model thus gives a satisfactory explanation for the origin of ledges of the type postulated by McLean.[13]

TABLE 2

Densely Packed Plane in the Coincidence-Site Lattice

Σ	Twin system	Densely packed plane	
		BCC	FCC
3	111, 112	112	111
5	012, 013	013	012
7	123	123	123
9	122, 114	114	122
11	113, 233	332	113
13_a	320, 015	015	320
13_b	134	134	134
15	125	125	125
17_a	140, 350	350	140
17_b	223, 334	334	223
19_a	133, 116	116	133
19_b	235	235	235

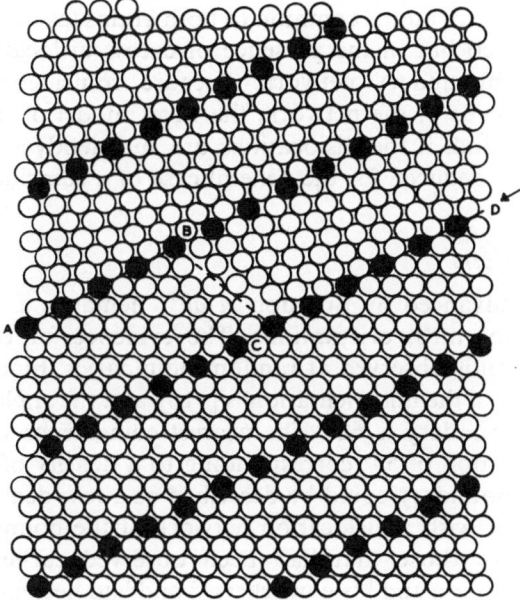

Fig. 9.7. Two-dimensional model for a b.c.c. bicrystal with a coincidence-site density of 1 : 11.

It is obvious that perfect coincidence is destroyed even with the least deviation of axis or angle from that required for a special relationship to exist between the two crystals. Field-ion micrographs of grain boundaries which do not possess coincidence-site relationships also show fit and misfit regions. This can only be explained if the coincidence model can retain some significance even for general high-angle boundaries. It is also known from the work of Aust and Rutter[14] that certain special characteristics of boundaries persist over a small range of misorientations around that of the special boundary.

All the above observations can be explained if the random boundary is considered composed of a special boundary and an associated sub-boundary to account for the angular (or axial, which are equivalent) deviation. The situation is analogous to a low-angle boundary on real lattice planes at a rotation of 0°. In the random boundary the dislocations are partial dislocations in the coincidence lattice. The preservation of the identity of dislocations forming the superposed subboundary requires that only small angular deviations of about 6° from the orientation relationship characteristic of a special boundary can be explained in this manner. It is seen from Table 1 that special boundaries up to a significant coincidence and an angular spread of 6° offer reasonable coverage of the orientations.

The random boundary thus differs from the special boundary not so much in the absence of atoms common to both grains (as would be expected from a rigid model) but in the presence of excess dislocations forming a superimposed subboundary network. Because coincidence regions still occur within the cells of the network, the Mott model of the boundary is still a good one, and, if the dislocations are sufficiently widely spaced, a ledge structure, as deduced for the general case of a coincidence boundary, will still exist.

9.6.2   Observations

The model advanced by Brandon et al.[11] demands a structure based on dislocations for most orientations. The evidence for the dislocations rested on dark bands observed at random boundaries. Figure 9.8 shows a bicrystal. The axis of the misorientation is [110], and the angle of misorientation across the boundary is 62°. The nearest coincidence-site relationship is for a misorientation of 59° corresponding to a density of coincidence of 1 : 33. Thus this boundary could be described as a low-density coincidence boundary together with a dislocation network at the boundary. The dark band is similar to those that had been observed earlier at low-angle boundaries by Brandon and his coworkers[15] and had been attributed earlier to the effect of preferential evaporation occurring as a result of the strain field of the dislocations. Further observations[16] have shown that this effect

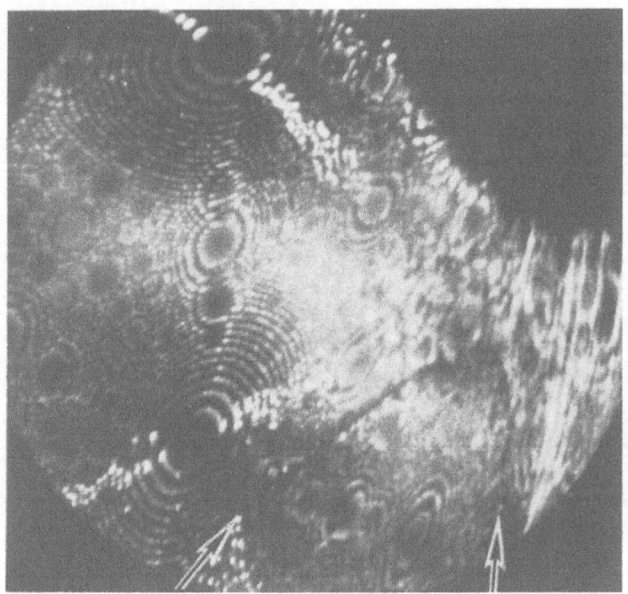

Fig. 9.8. A bicrystal in tungsten with the boundary appearing as a dark band.

does not occur at all low-angle boundaries. It appears that electronic effects can play an important part in the image contrast at grain boundaries and can lead to the appearance of streaks and dark bands. This has been discussed by Ranganathan *et al.*[17] (See also Section 8.5.2.) In the present section direct evidence for the presence of dislocations at grain boundaries will be given and then analyzed.

Figure 9.9 is a micrograph from a tungsten specimen containing a 50°-tilt boundary. The axis of misorientation is [110], and the plane of the boundary is (332). An accurate determination of the orientation relationships was possible in this case as the tip was symmetrical and [110] was at the center of the micrograph. The relationship corresponds to a lattice coincidence between the two grains with a density of coincidence sites of 1 : 11. The two-dimensional coincidence-site model given in Fig. 9.7 corresponds to the same rotation about the same axis. The boundary lies very nearly along the most densely packed plane in the coincidence-site lattice, where all the atoms are located on coincidence sites. There are a few ledges of about 10-Å dimension to be seen in Fig. 9.9. The atomic fit at the boundary is excellent, and there is no evidence of preferential evaporation.

Figure 9.10 shows a grain boundary in molybdenum. The misorientation is around the [110] axis and is 30°. The relationship is 3° away from

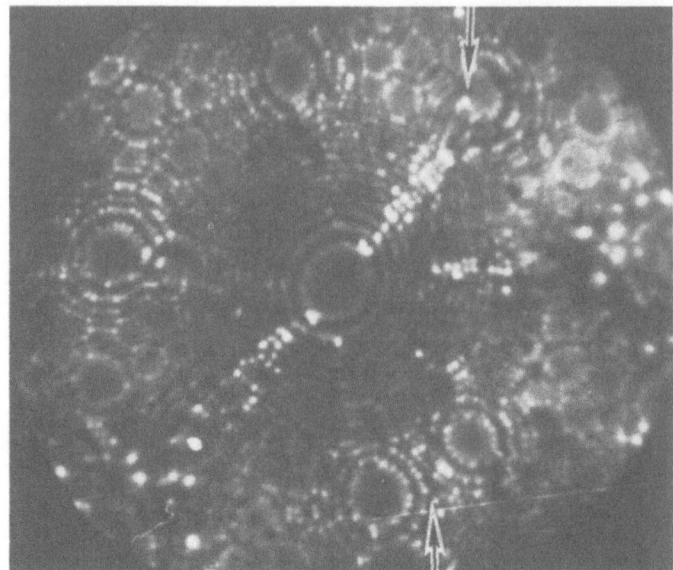

Fig. 9.9. A special boundary in tungsten with a coincidence-site density of 1 : 11.

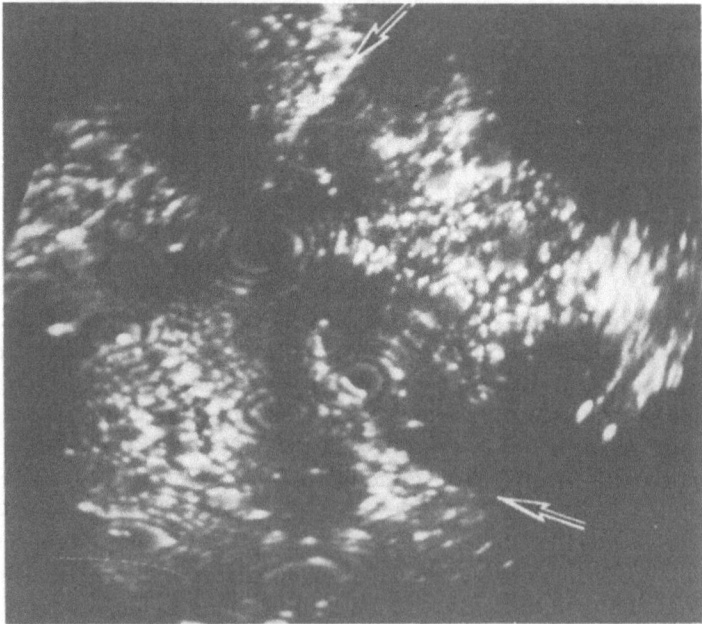

Fig. 9.10. A random boundary in molybdenum. The boundary is composed of a special boundary and a small-angle boundary. The spiral effect has its origin in the latter effect.

[110] $-$ 27°, which represents a special boundary with a density of coincidence of 1 : 19. One can then consider the boundary to be made up of this special boundary and a 3°-small-angle boundary. The spiral effect seen in the (110) planes has its origin in the latter boundary. The boundary plane is a few degrees off $(1\bar{1}0)_B$ and $(3\bar{3}2)_A$ and is near a densely packed coincidence-site lattice plane of the [331] type in both crystals. This accounts for the atomic matching across the boundary.

In the iridium example the dislocation formed part of a low-angle boundary; in the case of tungsten it accounted for the deviation of the grain boundary from a close-packed plane. Now in molybdenum a similar spiral forms part of the small-angle boundary superposed on a special boundary.

Figure 9.11 shows a tricrystal in iridium. The boundary B–C corresponds to a 6° misorientation around [111]. The boundary appears very nearly as a streak. Most (111) planes are seen to join smoothly across the boundary. Boundary A–B is a high-angle boundary with a misorientation of 52° around [111]. The nearest coincidence-site lattice is 5° away. The boundary is quite narrow and shows extremely good fit over nearly a hundred layers. The good fit observed at the boundary A–B merits some discussion. The

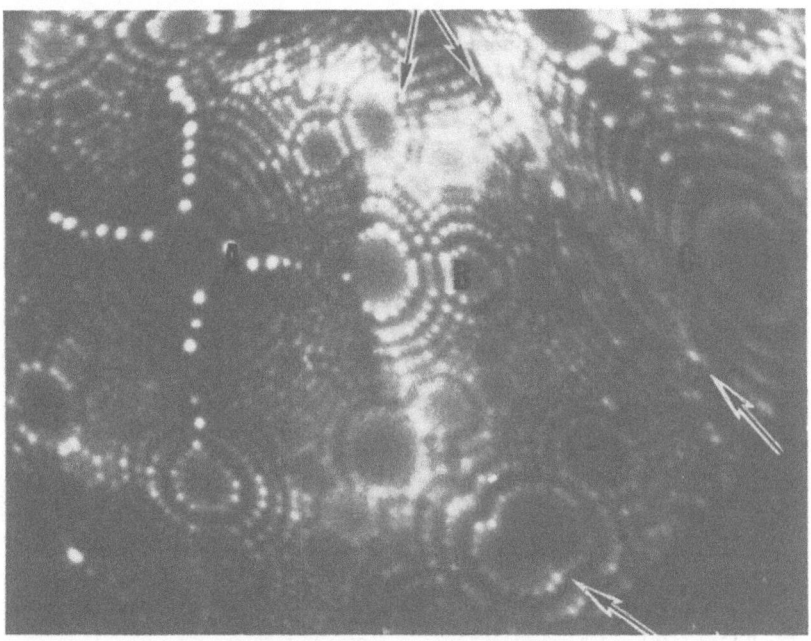

Fig. 9.11. Tricrystal in iridium.

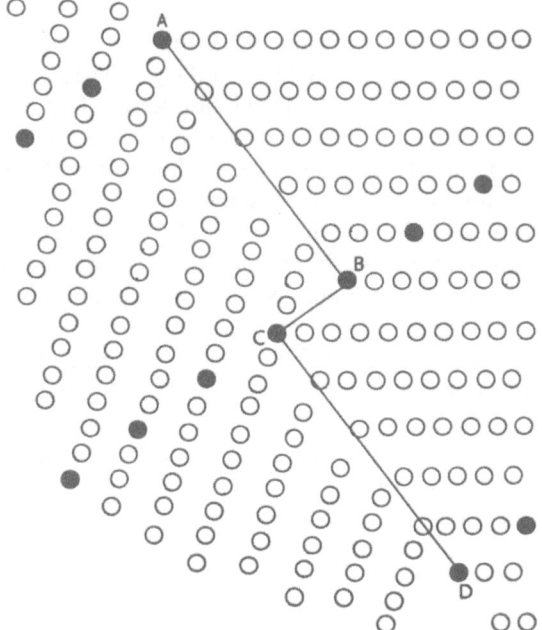

Fig. 9.12. Coincidence-site model for the high-angle boundary in the tricrystal shown in Fig. 9.11.

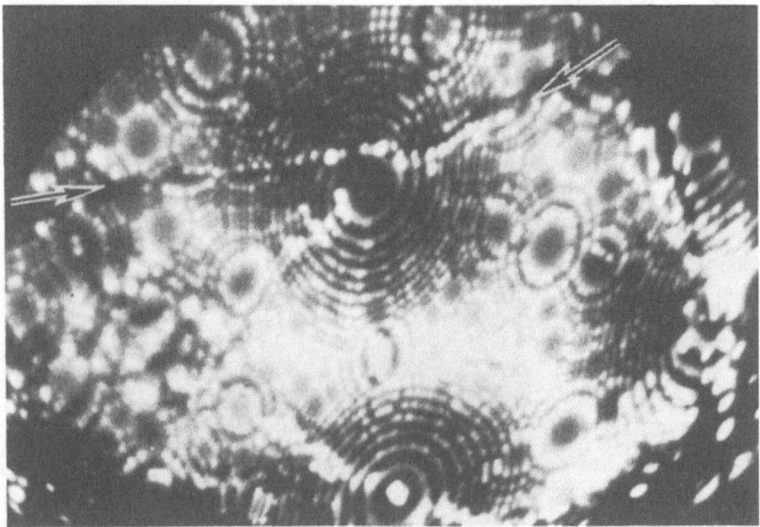

Fig. 9.13. A [110] — 28° boundary in tungsten. The deviation from lattice coincidence is slight in this case.

orientation relationship for this boundary is 5° away from $[111] - 47°$, which is the relationship for a coincidence-site lattice of $\Sigma = 19$. This lattice is illustrated in Fig. 9.12. It can be seen that the density of coincidence is so low in this case that an associated 5° subboundary will take out almost all lattice coincidence.

Figure 9.13 shows a $[110] - 28°$ boundary in tungsten. There appears to be a dark band associated with the boundary. On field evaporation this band became less prominent as the intersection of the boundary moved inwards owing to the inclination of the plane of the boundary to the axis of the tip. One can see a series of interlocked spirals. In some regions the continuation of the (110) planes across the boundary can be followed through a number of planes.

9.6.3 Discussion

Field-ion micrographs of high-angle grain boundaries show three types of structures. In one kind, the planes continue undisturbed across the boundary. Such structures have been observed in twins[4,7] as well as in a (100)-32° boundary in tungsten.[10] It is possible, however, that these boundaries can contain dislocations which have their Burgers vector parallel to the undisturbed planes. In the second kind, spiral structures, which can generally be associated with dislocations, have been observed. Most boundaries generally show a greater degree of disturbance at the boundary. Even in such cases the boundary shows good fit over a few regions. The conclusions to be drawn are that all the observations support the transition lattice theory, which states that the disturbance is confined to a few atom diameters near the boundary and that the two lattices continue right up to the boundary.[5] It can further be inferred that the boundaries consist of regions of fit and misfit. Mott[18] originally proposed such a model. The coincidence-site lattice model of Brandon et al.[11] is an attempt to characterize the regions of fit and misfit in crystallographic terms. Section 9.6.2 gave the evidence available for such a model from field-ion micrographs. (Brandon[19] has recently extended the theoretical part of the model in a rigorous way.)

9.7 Properties of Grain Boundaries

Every field-ion micrograph of a bicrystal can in principle be used to draw some information concerning atomic configuration at grain boundaries. That it also contains interesting details about the properties of grain boundaries is often a matter of chance. The small field of view and the loss in image intensity and resolution at high temperatures act as severe limitations.

Both thermally activated and stress-induced migration of grain boundaries have been observed by the author.[20] It is however difficult to throw any light on the atomistics of the process, as the image disappears at high temperatures and as the stress-induced migration occurs in an unpredictable manner.

Segregation of oxide at the boundary has been observed in boundaries in tungsten–5% rhenium.[11] Large bright spots are occasionally observed at boundaries. Their chemical identity is unknown. Investigation of segregation and depletion near grain boundaries is a fruitful field where very little has been done until now.

Grain boundaries in general increase the strength of a material, as they prevent the easy passage of dislocations. Pile-up of dislocations has been shown to occur at grain boundaries. Li[21] has postulated that, under the influence of stress, ledges in grain boundaries can be made to emit dislocations. Müller[22] has shown that grain boundaries in rhenium hardened the material sufficiently to prevent slip under imaging conditions. Ranganathan and Cottrell[16] have reported a particularly interesting interaction between a dislocation and a grain boundary in iridium.

A number of field-ion microscope studies are now directed toward a study of thin-film formation through evaporation of metals. These could be modified so that, in a bicrystal, diffusion along a grain boundary can be investigated.

The field-ion microscope lends itself readily to the study of small-scale effects arising from radiation damage. Grain boundaries interact with such imperfections in a marked way. The elegant studies of Barnes et al.[23] have shown that the ability of grain boundaries to act as sinks and sources for vacancies is important. Bowkett et al.[24] have observed a large vacancy cluster about 15 Å away from a grain boundary in neutron-irradiated tungsten. The author[20] has carried out a preliminary study of helium bubble formation in tungsten and has observed an "incipient" bubble near a low-angle grain boundary.

One fruitful line of investigation would have been the study of specificity of adsorption at grain boundaries. Ehrlich and Hudda[25] have shown that adsorbed atoms can be imaged with the field-ion microscope. Brenner[26] carried out experiments to determine whether there is enhanced binding of adsorbed gas atoms at grain boundaries. He found that adsorption seemed to occur preferentially at the edge of the (110) planes but not at the boundary.

9.8   Interphase Interfaces

The atomic configuration of an interface separating two crystals which differ in composition and structure presents problems similar to that of the atomic configuration at high-angle grain boundaries. However very few studies have been carried out. Ralph and Brandon[3] have observed that a

σ-phase precipitate particle in the 34% rhenium alloy gave enhanced contrast and that in spite of the lack of crystalline regularity in the disordered matrix the partial coherency with the matrix was readily detectable. The precipitate did not show many developed crystal planes. Recently McLane and Müller have observed the NiBe–Nickel solid solution interface. The precipitate has a CsCl structure, while the matrix (98 at.% Ni − 2 at.% Be) has a f.c.c. cubic structure. They also showed a second type of interface across which complete coherence was observed. They tentatively interpreted it as a Guinier–Preston zone. Müller et al.[28] have also reported the observation of a lens-shaped martensite plate in a high-carbon steel. While the identification is only tentative, these examples show that much of the promise of the technique is turning into reality.

Acknowledgments

Much of the work reported here formed part of the author's doctorate thesis. The author is grateful to the United Kingdom Atomic Energy Authority for the provision of research facilities. It is a pleasure to thank Professor A. H. Cottrell, Dr. D. G. Brandon, and other members of the field-ion group in Cambridge for many stimulating discussions. The author is also grateful to Professor G. Thomas for encouragement and the United States Atomic Energy Commission for financial support.

References

1. E. G. Brock, *J. Appl. Phys.* **28**: 241 (1957).
2. P. Wolf, *6th Field-Emission Symposium*, Washington, D.C., 1959.
3. B. Ralph and D. G. Brandon, *Phil. Mag.* **8**: 919 (1963).
4. J. Hren, *Acta Met.* **13**: 479 (1965).
5. S. Amelinckx and W. Dekeyser, *Solid State Physics* **8**: 325 (1959).
6. E. W. Müller, *Acta Met.* **6**: 620 (1958).
7. E. W. Müller and O. Nishikawa, *Technical Report AFML-TR*-65-201, Aug. 1965.
8. E. O. Hall, *Twinning*, Butterworth & Co. (London), 1954.
9. E. Votava and A. Berghezan, *Acta Met.* **7**: 392 (1959).
10. H. F. Ryan and J. Suiter, *Acta Met.* **14**: 847 (1966).
11. D. G. Brandon, B. Ralph, S. Ranganathan, and M. Wald, *Acta Met.* **12**: 813 (1964).
12. S. Ranganathan, *Acta Cryst.* **21**: 197 (1966).
13. D. McLean, *J. Australian Inst. Metals* **8**: 45 (1963).
14. K. T. Aust and J. W. Rutter, *Recovery and Recrystallization of Metals*, L. Himmel, ed., Wiley (Interscience) (New York), 1963.
15. D. G. Brandon, M. Wald, M. J. Southon, and B. Ralph, *J. Phys. Soc. Japan*, **18**: 324 (1963).
16. S. Ranganathan and A. H. Cottrell, *Third Regional Conference on Electron Miscroscopy*, Czech. Acad. of Sciences, Prague, 1964, p. 163.
17. S. Ranganathan, K. M. Bowkett, J. Hren, and B. Ralph, *Phil. Mag.* **12**: 841 (1965).
18. N. F. Mott, *Proc. Phys. Soc. (London)* **60**: 394 (1948).
19. D. G. Brandon, *Acta Met.* **14**: 1479 (1966).
20. S. Ranganathan, Ph.D. Thesis, Cambridge University, 1965.
21. J. C. M. Li, *J. Australian Inst. Metals* **8**: 206 (1963).
22. E. W. Müller, *Third Regional Conference on Electron Microscopy, Czech. Acad. of Sciences, Prague*, 1964, p. 161.

23. R. S. Barnes, G. B. Redding, and A. H. Cottrell, *Phil. Mag.* **3**: 97 (1958).
24. K. M. Bowkett, J. Hren, and B. Ralph, *Third Regional Conference on Electron Microscopy, Czech. Acad. of Sciences, Prague,* 1964, p. 191.
25. G. Ehrlich and F. G. Hudda, *J. Chem. Phys.* **36**: 3233 (1962).
26. S. S. Brenner, "Properties of Solid Surfaces," *Joint Symposium ASM-AIME, New York,* Oct. 27–28, 1962.
27. S. B. McLane and E. W. Müller, *12th Field-Emission Symposium,* Pennsylvania State University, University Park, Pa., 1965.
28. E. W. Müller, S. Nakamura, O. Nishikawa, and S. B. McLane, *J. Appl. Phys.* **36**: 2496 (1965).

Chapter 10

EXPERIMENTAL STUDIES OF ALLOYS WITH FIELD-ION MICROSCOPE

B. Ralph†

10.1 Introduction

The field-ion microscope has been developed to a point that almost routine studies of some phenomena in metals are possible. A considerable amount of effort has been directed by four groups (those of Professor Müller, Professor Machlin, Professor Krautz, and the Cambridge group) toward applying the field-ion microscope to studies of phenomena occurring in binary alloys. For some types of alloy studies the field-ion microscope has proved largely unsuccessful so far, while in other areas the full potentialities of the field-ion microscope have been realized. It is the purpose of this review to summarize the difficulties encountered in these studies and to indicate those alloy studies which can profitably be made and where new information may be derived.

Müller published the first account of his alloy studies in 1961[1] and reviewed his alloy work again in 1962 at the Kyoto conference.[2] At that same meeting the first account of the Cambridge group's work in this area was also presented.[3] This was followed by a detailed presentation of studies of three tungsten–rhenium alloys,[4] and Caspary and Krautz have made a detailed study of a series of tungsten–molybdenum alloys.[5]

At this early stage it was apparent that studies of finite-concentration solid-solution alloys were difficult in that the images obtained were extremely irregular and interpretation, which relies on easy identification of crystallographic features, was extremely difficult and largely qualitative. It is likely that the computor simulation of field-ion patterns is the most suitable method for direct interpretation (see Chapter 5).

† Lecturer in Metallurgy, University of Cambridge, Cambridge, England.

To date the most successful studies have fallen into 2 classifications:

1. Order–disorder studies
2. Studies of segregation phenomena

These two types of investigation will be reviewed after a discussion of image formation from alloys.

10.2 Image Formation from Alloys

The presence of solute atoms is expected to affect both the field ionization process and the field evaporation process. The effect of the presence of a solute atom on the field ionization process can be seen most clearly by examining field-ion micrographs of dilute solid-solution alloys where in some cases differences in image spot intensity can be attributed to the presence of a solvent atom. For instance, at low concentrations of rhenium in tungsten some dim image points are seen and the number can be approximately related to the atomic concentration of rhenium present. An analysis based on this approach is difficult in that similarly situated atom sites have to be compared.

This effect can be understood in terms of the present theories of the field ionization process (see Chapter 2). The ionization probability above an atom is a very sensitive function of the field, and Müller predicts a 30% change in the ionization probability for field fluctuations of only 1%.[6] Such field fluctuations can be pictured as arising above solute atoms where a localized perturbation potential exists.

It is particularly instructive to consider the modification of the electron levels involved by introducing a solute atom into a pure metal since the image in a field-ion microscope is a distorted representation of the electronic structure. The electron theory of alloys can be summarized by saying that an alloy will assume that type of crystal structure in which there are enough low-energy states to accommodate all of the electrons in the crystal. Siverston and Nicholson[7] have discussed the principal factors which allow a description of the electronic structure of an alloy.

The addition of a solute atom to the solvent matrix produces a change in the electronic energy of the solid, ΔE. If the ionization energy of the solute atom is greater than that of the solvent, then the ion core of the solute will attract a valence electron more strongly than will a solvent atom. This will result in an accumulation of charge at the solute ion at the expense of the electronic charge distributed over the rest of the lattice. These charge *shifts* may result in either a *polarization* of the conduction band near the solute atom or the localization of some of the valence electrons in *bound states*. The opposite situation is expected if the ionization energy of the solvent is greater than that of the solute. The immediate effect in both cases is that the potential field will become smoothed out by the valence electron charge shifts.

The energy change ΔE is due to the modification of the electronic states by the addition of the solute. The charge density and therefore the electrostatic field of the crystal are no longer perfectly periodic. The solute ion produces a perturbing field V_p which will tend to modify the state of motion and energy of the valence electron. Slater and Koster[8] treated this problem by considering a crystal whose electrostatic potential is a periodic function except for a localized nonperiodic force. They were able to show that the solution to such a problem can be expressed in terms of the original functions of the unperturbed periodic lattice. They were also able to show the effect of a perturbing potential V_p, localized at an atomic site, on the allowed band structure. If V_p is less than a certain value, the energy levels are distributed in about the same manner as they are in the perfect pure solvent. If V_p is greater than this critical value, one of the energy levels forms a discrete state (called a *bound state*) by separating from the energy band. This bound state tends to be localized at the solute atom. In the case of a random distribution of solute atoms, then, the energy levels will be relatively unaffected; at most, only slight shifts will occur. For greater differences in nuclear charge discrete states will separate from the band structure.

Friedel[9] has considered the valency and size effects in a solid solution for both dilute and finite concentrations and by considering valency effects alone has shown that for dilute solutions the perturbation potential V_p can be expressed in terms of Z, the difference in chemical valency between the solute and solvent atoms

$$V_p = \frac{Z}{r} e^{-qr} \tag{10.1}$$

where

$$q^2 = 4\pi N_0(E_{max}) \tag{10.2}$$

In these equations $N_0(E_{max})$ is the density of states at the Fermi surface, r is the radial distance from the solute atom, and q is a measure of the screening radius; $q^{-1} \sim 1$ Å.

To determine Z, Friedel assumed that the valencies are the normal chemical ones with the exception of the transition elements. For transition elements he assumed that nickel, palladium, and platinum behave as if they were zero valent, while cobalt, rhodium, and iridium should have a valency of -1, iron, ruthenium, and ozmium, a valence of -2 etc.

For finite concentrations of solute atoms the solute interactions were considered to determine the radius R of the spheres of influence of the impurities. Thus R_0 was found to be less than five atom diameters for most metals, and for iron and nickel, the only transition metals tabulated, R_0 was found to be one atomic diameter. As R_0 is the radius of the sphere of total screening at infinite dilution, in alloys of transition metals the perturbations are expected to be completely localized. Friedel has further shown that, if polarization effects are neglected and only valence differences Z are considered, then for $Z < 0$ solute atoms will cluster, and for $Z > 0$ unlike neighbors are favored, i.e., the material tends to exhibit short-range order. This means that to consider the structure of solid solution as a random array of solute atoms in the solution is probably a considerable oversimplification.

This sort of approach qualitatively accounts for the type of field-ion patterns obtained from dilute solid solutions. Where the solute atoms are metallic, Friedel's approach suggests that the perturbation is strongly screened and hence any changes in electronic structure will be confined to the immediate vicinity of the solute atoms. This appears to be the case for a tungsten–5% rhenium alloy, where in fact a percentage of image spots are somewhat dimmer than their immediate neighbors. This model can also be stretched to account for the appearance of interstitial impurity atoms on field-ion micrographs. These impurities (such as C, N, and O) give rise to very bright image spots which are considerably larger than those from matrix atoms. Here Friedel would predict that the perturbation in the lattice would be greatest and not so closely screened.

In finite-concentration solid-solution alloys the image is invariably irregular which indicates an irregular mode of field evaporation. Even in dilute solid solutions preferential retention or evaporation occurs. Machlin and DuBroff[10] have looked at a series of dilute platinum alloys and correlated their behavior with the theory of field evaporation developed by Brandon.[11] (For a detailed discussion of field evaporation from alloys see Section 3.7.) Two comments arising out of the observed pattern of field evaporation seem pertinent to this discussion. Firstly, the degree of irregularity of any selected area depends critically on the amount of solute present, and, secondly, this degree of irregularity depends on any tendency the solute has to cluster and form a precipitate or to arrange itself on a superlattice and form a region of short-range order. Thus the observed nature of the image can be used to give a qualitative picture of concentration segregation effects and also to look at the development of short-range order.

10.3 Distinguishing between Atomic Species

In dilute solid-solution alloys a distinction can be made where preferential field evaporation or retention of solute occurs and also where there is a significant difference in image spot intensity. Such an analysis is relatively difficult and requires very careful and tedious examination of field-ion images.

In more concentrated solid-solution alloys an analysis by the above

method is impossible owing to image irregularity, but, as mentioned previously, the relative degree of image irregularity in any one crystallographic
region gives a qualitative picture of the local solute concentration.

A method for distinguishing between atomic species in some alloys has
been reported.[12] A plot of the number of image points seen in one crystallographic region as a function of tip voltage between the ionization threshold
field and the field for evaporation is approximately linear for pure metals but
shows a discontinuity, which gives the approximate atomic concentration
of the alloy, for a solid-solution alloy. An example of such a plot is given in
Fig. 10.1, which is from a tungsten–26% rhenium alloy specimen. The
general applicability of this method is uncertain, as is the exact reason why
this discontinuity occurs. What is quite clear is that, even for those alloys
where this method does work, without automation of the image recording
and spot counting such a technique is of only very limited use.

Where a distinct difference in image spot size or intensity exists between
solvent and solute, then the spot-counting method can prove accurate and
not particularly tedious. For instance, Fortes and Ralph[13] have found a

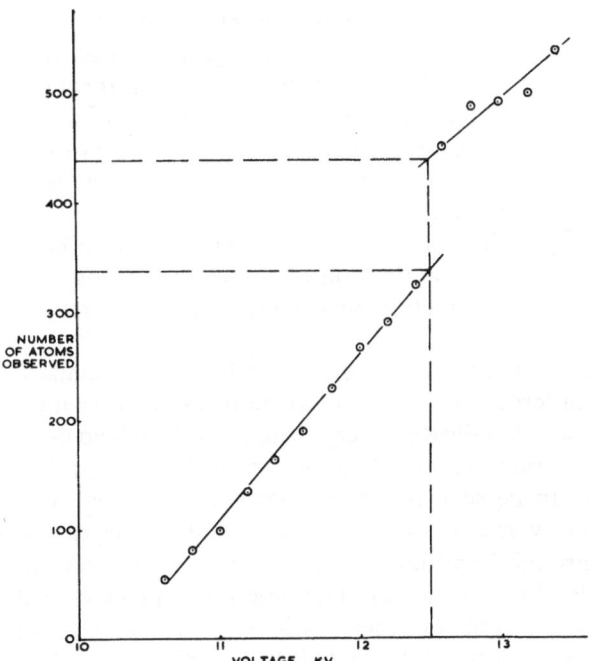

Fig. 10.1. Voltage dependence of the number of atoms observed
for a specimen of tungsten–26% rhenium alloy.[12]

correlation between the number of bright spots found on micrographs from iridium and the oxygen impurity concentration.

10.4 Order–Disorder Studies

Müller in 1962 showed images from an equiatomic cobalt–platinum alloy in the ordered and disordered states.[14] This work inspired the Cambridge group to begin a program studying this particular order–disorder reaction in detail with the field-ion microscope.[15,16] Further work on this system has also been performed by Tsong and Müller.[17]

Figure 10.2 shows a typical helium-ion image obtained from the disordered alloy. In this stage the alloy is equivalent to an ordinary solid-solution alloy with high (50%) solute content, and the image can be seen to be extremely irregular. In interpreting images from disordered alloys two factors warrant consideration. Firstly, the degree of development of crystal planes varies with the plane, even for an alloy which is uniformly disordered. In this micrograph the only planes identifiable as such are those of very low index. Preferential field evaporation of one species obscures the development of the higher-index planes. Secondly, it has been found in practice that Pt–Co is not uniformly disordered but contains small regions of high local order which tend to lead to a regular development of planes in the area where the surface intersects them. In the region marked A a well-defined ring structure can be seen; however field evaporation through a large number of atomic layers discloses that this regularity is not a crystallographic feature but is in fact due to a small region of high local order in the bulk specimen. After field evaporation through this region the well-defined ring structure is no longer seen.

Specimens in various states of partial order have also been studied, and it is possible by careful examination of successive micrographs from specimens to make a clear distinction between the relative degree of order in local regions.

In Figure 10.3 the high degree of crystallographic regularity obtainable from the fully ordered state ($CoPt_1$) is demonstrated. Very high index planes are developed, which indicates a very regular mode of field evaporation. This regular mode characterizes a lattice on which the two species are regularly distributed. It can be seen that the degree of perfection developed by the individual planes varies to an appreciable extent. In this case, however, field evaporation through hundreds of atomic layers shows that this irregularity is characteristic of the particular crystallographic plane considered and not of a local region of the specimen. Although for a perfect superlattice the atomic arrangement on any plane will be completely regular, the distribution of atoms on the plane and the stacking sequence of these planes will characterize the mode of field evaporation. A complicated arrangement will

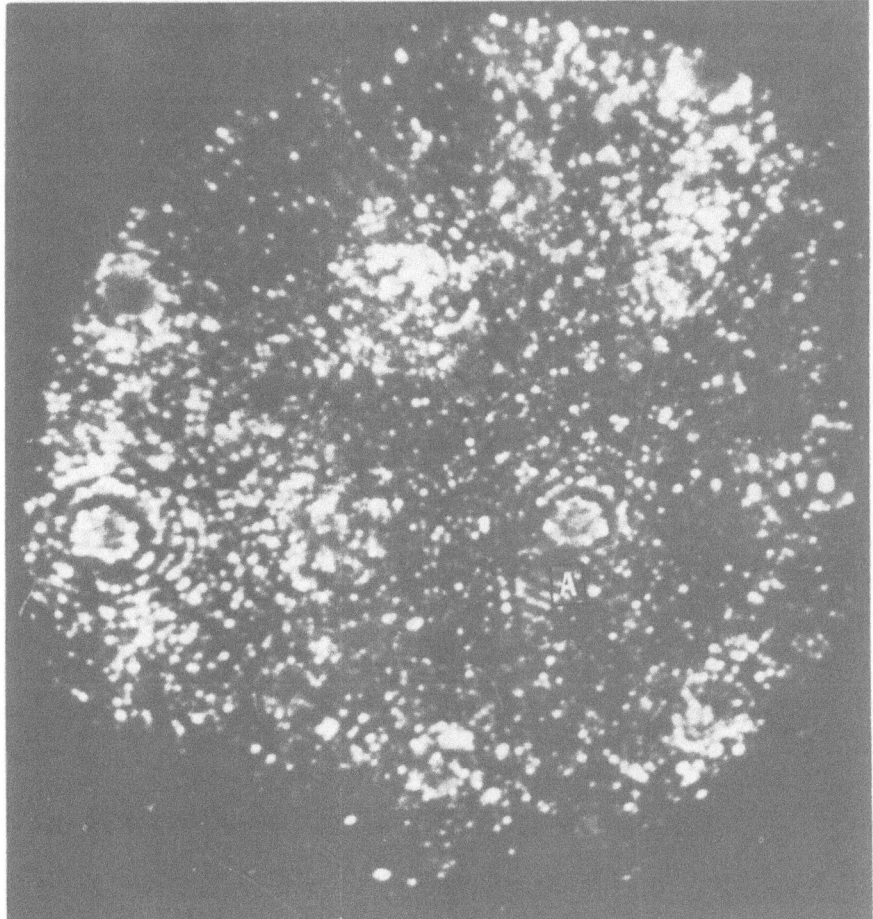

Fig. 10.2. Disordered equiatomic cobalt–platinum alloy. In the region marked A well-developed lattice plane rings may be seen.[16]

reduce the specificity of binding energy for an atom on the plane, and preferential field evaporation may occur.

A geometrical model of the distribution of atoms on a crystallographic plane and the stacking sequence of these plane layers can account for the relative development of each crystallographic area. In principle this relative development can be correlated with the binding energy of the superlattice. Such a correlation is being attempted, although it is difficult owing to the approximately 10% lattice expansion occurring in the specimen during image formation.

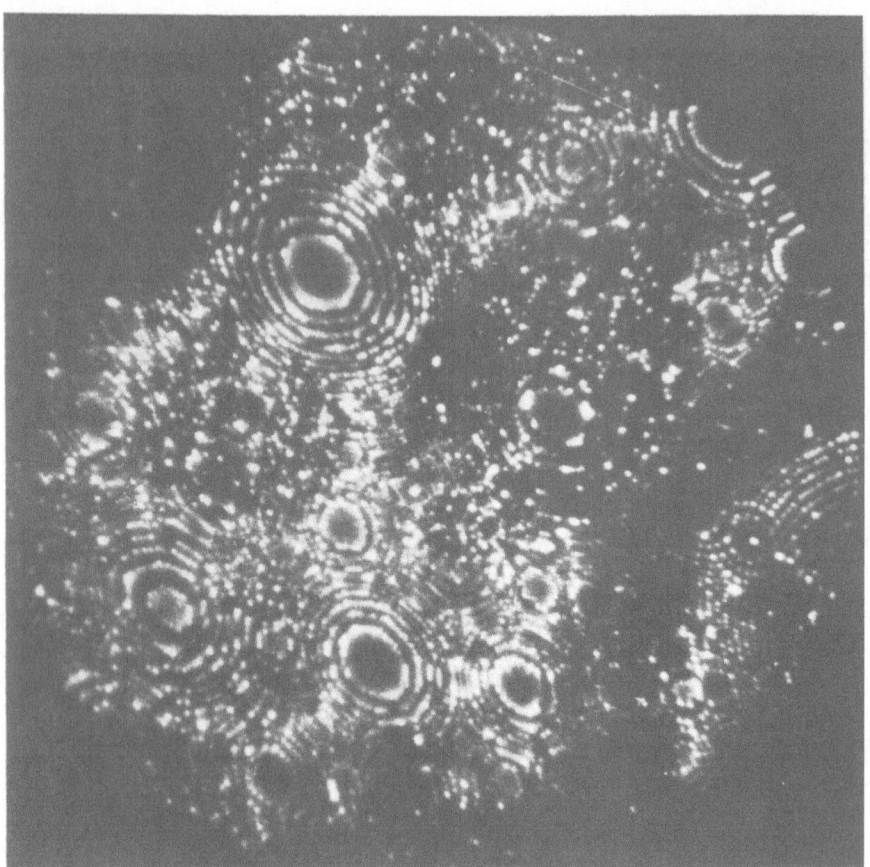

Fig. 10.3. Ordered equiatomic cobalt–platinum.[16]

One feature of these ordered patterns of particular interest is the pre-dominance of the (001) plane region. A micrograph from a specimen with an {001} axis is included as Fig. 10.4. The step height on the (001) plane was calculated by counting the number of steps between two poles of known angular separation and by using a value of the tip radius calculated from a knowledge of the applied tip voltage and the threshold ionization field. This step height turned out to be the same as the c parameter and hence it can be seen that two atomic layers, alternately all platinum and all cobalt, comprise the step, since the $CoPt_1$ structure is comprised of alternate (002) layers of all cobalt and all platinum. A schematic cross section of such a specimen is given in Fig. 10.5. When the topmost double (002) layer has been field evaporated down to a critical size, the two layers then evaporate singly,

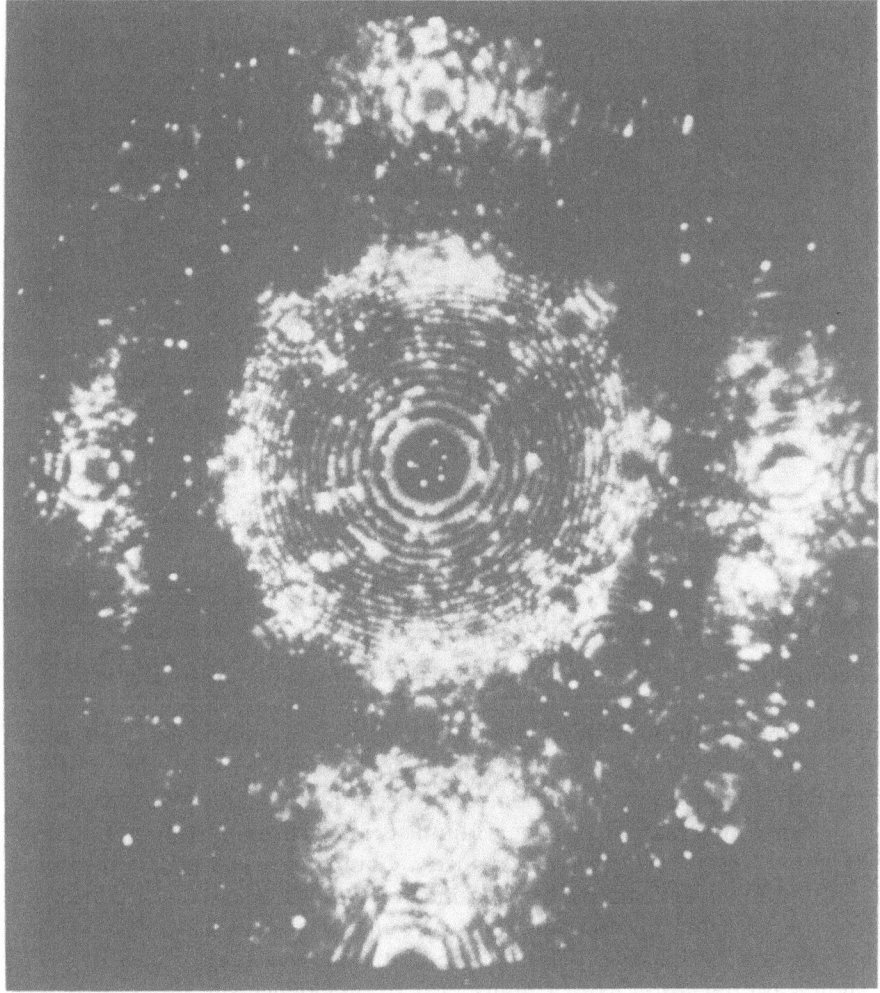

Fig. 10.4. Ordered cobalt–platinum in the (001) orientation showing the dominance of the (001) plane.[16]

an occurrence that should aid the correlation of observed end-form structure with the superlattice energy.

It is clear that the field-ion microscope is a proven technique for studying short-range and long-range order. Lattice defects such as translational and rotational antiphase domain boundaries and atoms misplaced on the wrong sites in the superlattice can be identified and characterized.[16] Recently, Newman *et al.* have obtained micrographs from Ni_4Mo and

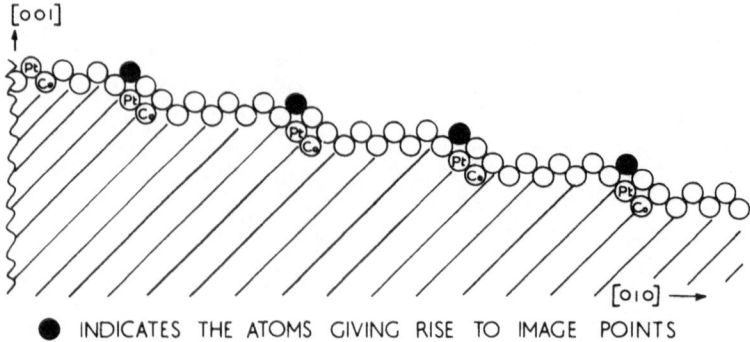

Fig. 10.5. Radial section through the specimen tip showing the alternate layers of Pt and Co atoms forming the double step.[16]

examined defects in this lattice.[18] An example of fully ordered Ni_4Mo is given in Fig. 10.6.

10.5 Segregation Studies

A recent study of the segregation of interstitial oxygen impurities at grain boundaries in iridium has demonstrated the potential of the field-ion microscope for such studies.[13] Previously examples have been given of the condensed segregation of oxygen at grain boundaries in W–5% Re[4] and of segregation of Re toward boundaries in W-Re alloys.[19]

The starting material for the iridium studies is iridium containing about 500 ppm of oxygen as the main impurity. Counts of bright spots on iridium micrographs, away from the bright zone line decoration, give concentrations in agreement with this 500 ppm. Grain boundaries which have coincidence-site relations (see Chapter 9) show little in the way of segregation, but, where a significant deviation from a high-density coincidence-site exists, a pattern of bright spot segregation to the boundary region is seen after various annealing treatments.

Figure 10.7 is a micrograph from an iridium specimen which has been annealed at 1200°C. A crystallographic analysis of this grain boundary over many atom layers shows that the axis-to-angle relationships is [111]/56° $\pm$ 2°. Thus the rotation deviates by about 4° from the perfect twin relationship. A pattern of bright spots can be seen toward this grain boundary, and a histogram showing the segregation pattern found over many layers of field evaporation is shown in Fig. 10.8. Evidence of segregation of oxygen to dislocations in iridium has also been found.[13]

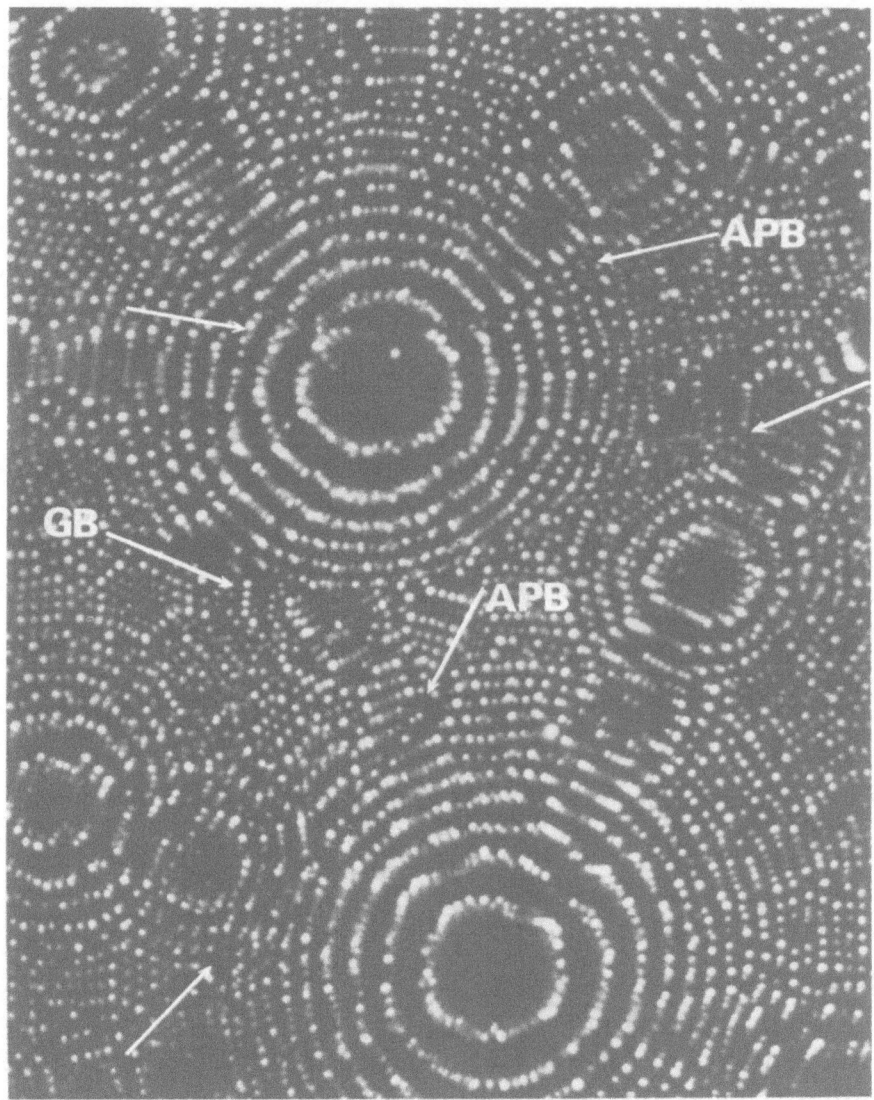

Fig. 10.6. Several antiphase boundaries and a grain boundary in fully ordered Ni_4Mo.[18]

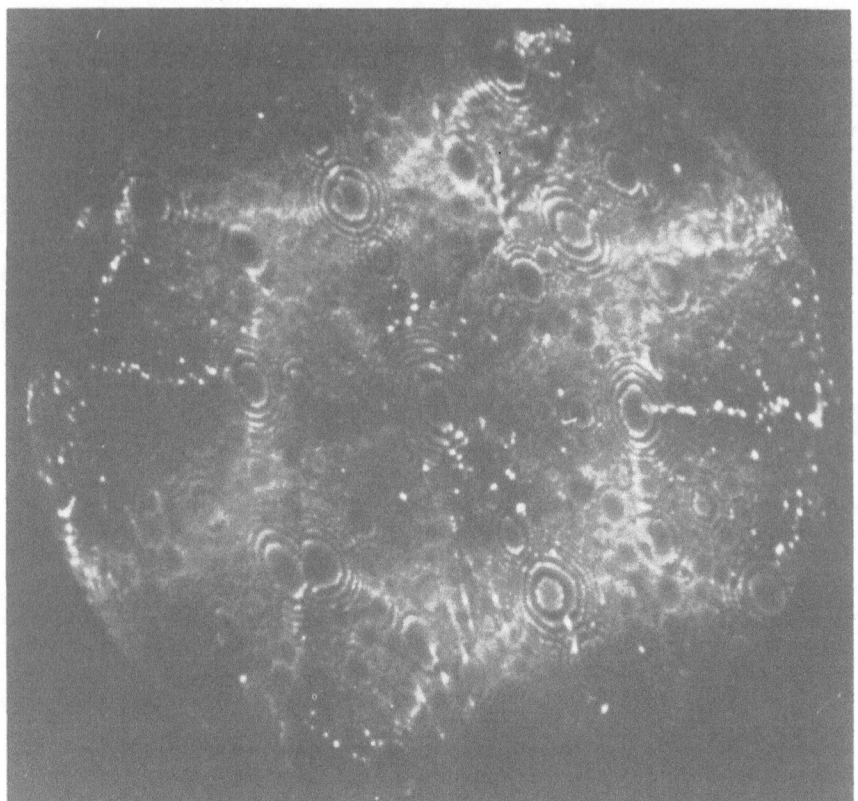

Fig. 10.7. A grain boundary in iridium showing segregation of bright spots at the boundary region.[13]

10.6 Discussion

Improvements in our understanding of the parameters affecting the imaging process from alloys allow a wider and wider range of studies to be envisaged. Some two-phase materials have already been studied, e.g., σ-phase precipitates in the β matrix of a W–34% Re alloy[4] and more recently pearlite in a plain carbon steel.[20] The study of two-phase materials with the field-ion microscope is at an early stage, but the combined use of field-ion and electron microscopy in such studies constitutes an important development in this work.[20] As yet the possibility of studying phenomena in ternary alloys seems distant.

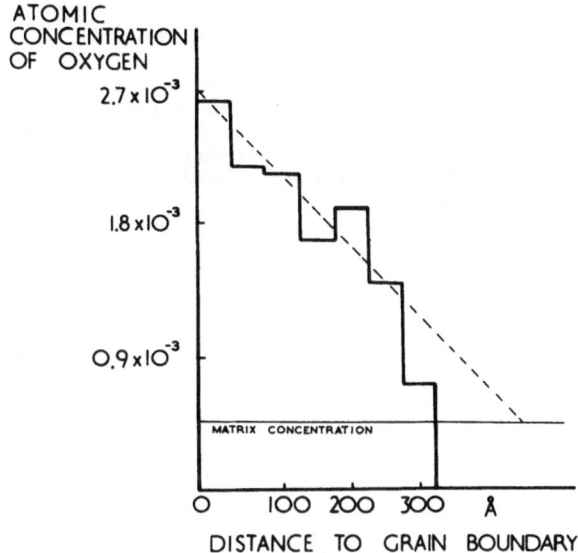

Fig. 10.8. Histogram showing the segregation of oxygen at the grain boundary, illustrated in Fig. 10.7.[13]

Acknowledgments

The work of the Cambridge field-ion group is supported by the United Kingdom Atomic Energy Authority, Harwell, and the Science Research Council (U.K.) to whom the author is grateful for encouragement and permission to present this account. Professor A. H. Cottrell, Fellow of the Royal Society, has been most generous with helpful discussions and encouragement.

The author is particularly grateful to Mr. H. N. Southworth, Mr. M. A. Fortes and Dr. R. Morgan for help in the preparation of this article and permission to quote their work freely.

References

1. E. W. Müller, *Imperfections in Crystals*, Wiley, (Interscience) (New York), 1961, p. 77.
2. E. W. Müller, *J. Phys. Soc. Japan* **18** *Suppl.* II: 1 (1963).
3. D. G. Brandon, M. Wald, M. J. Southon, and B. Ralph, *J. Phys. Soc. Japan* **18** *Suppl.* II: 324 (1963).
4. B. Ralph and D. G. Brandon, *Phil. Mag.* **8**: 919 (1963).
5. E. K. Caspary and E. Krautz, *Z. Naturforsch.* **19**: 591 (1964).
6. E. W. Müller, *Advan. Electron. Electron Phys.* **13**: 83 (1960).
7. J. M. Siverston and M. E. Nicholson, *Prog. Mater. Sci.* **9**: 303 (1961).
8. J. Slater and G. Koster, *Phys. Rev.* **94**: 139; **95**: 1167; and **96**: 1203 (1954).
9. J. Friedel, *Advan. Phys.* **3**: 446 (1954).
10. E. S. Machlin and W. duBroff, *12th Field-Emission Symposium*, Pennsylvania State University, University Park, Pa., 1965.

11. D. G. Brandon, *Surface Science* **3**: 1 (1965).
12. B. Ralph and D. G. Brandon, *J. Roy. Microscop. Soc.* **82**: 179 (1964).
13. M. A. Fortes and B. Ralph, *Acta Met.* **15**: 707 (1967).
14. E. W. Müller, *9th Field-Emission Symposium*, Williamstown, Mass., 1962.
15. B. Ralph and D. G. Brandon, *Journées Intern. Applications Cobalt* (Bruxelles) (1964).
16. H. N. Southworth and B. Ralph, *Phil. Mag.* **14**: 383 (1966).
17. T. T. Tsong and E. W. Müller, *J. Appl. Phys.* **9**: 7 (1966).
18. R. W. Newman, B. G. Le Fevre, and J. J. Hren, *13th Field-Emission Symposium*, Cornell University, Ithaca, N.Y., 1966.
19. B. Ralph, Ph.D. Thesis, Cambridge University, 1964.
20. R. Morgan, R. G. Faulkner, and B. Ralph, *J. Iron Steel Inst.* (*London*) **204**: 943 (1966).

Chapter 11

FIELD-ION MICROSCOPE STUDIES OF RADIATION DAMAGE

B. Ralph†

11.1 Introduction

This review is divided into two parts since the radiation-damage experiments performed thus far with the field-ion microscope naturally fall into two classifications. The first part (Section 11.2) covers field-ion microscope studies of neutron damage and the annealing of this damage, while part two (Section 11.3) is concerned with all other radiation-damage experiments in which the specimen is prepared in a field-ion form prior to irradiation.

11.2 Neutron Damage and Annealing Studies

The importance of understanding the mode of interaction of neutrons with crystals is paramount in any consideration of the economic viability of nuclear power. The properties that are altered by the interaction of radiation with crystals are well documented.[1–4] For metal crystals the most important property changes occurring during irradiation are in mechanical properties. Changes in susceptibility to brittle failure are especially important.

It is well known that these mechanical property changes arise from the interaction between glissile dislocations and the debris produced by irradiation. This debris is largely in the form of single-point defects and small clusters of defects. In some materials the largest of the defect clusters may be resolved in the electron microscope, but field-ion microscopy is the only technique allowing a direct resolution of the smaller defect clusters and the individual point defects as well. It should be emphasized at this point that one aim of any field-ion study of neutron damage is to obtain a correlation between the defect distribution found by field-ion microscopy and that inferred from property measurements or computer simulations.

† Lecturer in Metallurgy, University of Cambridge, Cambridge, England.

Of considerable theoretical and practical interest are the annealing characteristics of neutron damage. It is found experimentally that the damage is removed (at a maximum rate) in distinct temperature stages and substages (normally electrical resistivity is used to determine the exact kinetics of this removal). Five distinct stages are characterized with a large conflict around which species are migrating in stages 3 and 4. It can reasonably be expected that the field-ion microscope will help to resolve this conflict.

This presentation concentrates primarily on a description of the experimental conditions and techniques used in studies at present under way at Cambridge. In conclusion a brief description of the present state of these investigations will be given.

11.2.1 Experimental: Specimen Material

In principle, radiation-damage studies may be made on any material which gives stable field-ion micrographs. In practice large numbers of specimens of any one material have to be studied and some thousands of micrographs taken in order to get results which are statistically significant. Further, it is necessary to obtain images which are as regular as possible so that individual point defects may be identified. These limitations restrict such a study to refractory metals at the present time, and it seems likely that, once the nature of the damage in one particular b.c.c. metal and in one f.c.c. metal is established, the pattern of damage in other b.c.c. and f.c.c. metals may be obtained by correlating field-ion evidence, obtained from just a few specimens of these other metals, with that obtained from detailed refractory-metal studies.

It is vitally important to have the metallurgical state of the starting material well-defined. In practice superpure metals are used since any impurities present are known to influence the pattern and annealing characteristics of the damage. (This point is of particular importance concerning the presence of interstitial impurities in b.c.c. metals.) It also assists interpretation if the starting material is annealed to remove cold work prior to irradiation.

11.2.2 Irradiation Techniques

Short lengths of wire are irradiated prior to specimen preparation for field-ion microscope examination. This then precludes irradiations' being performed in a cryostat, and the pattern of damage seen corresponds to that obtained at the reactor ambient temperature (which is usually 60 to 150°C). In principle it would be desirable to carry out the irradiations at a lower temperature so that the low-temperature annealing stages (i.e., 1 and 2) might be investigated. In practice this would require specimen preparation prior to irradiation, and the pattern of damage would then not be expected to represent the bulk pattern owing to the influence of the large surface to volume ratio.

It is important to choose the total irradiation dose and energy of the neutrons so that within the volume of material examined in any one field-ion experiment there is a statistically good chance of finding a small number of primary events. Only fast neutrons with energies greater than about 1 MeV will produce primary displacements in the refractory metals, and, to avoid problems associated with transmutation caused by neutrons, it is often convenient to encapsulate specimens in cadmium or boron, which absorb the thermal neutrons.

The concentration of primary events, C_p, formed by irradiation in a fast neutron flux ϕ for a time t is given by

$$C_p \sim \phi t \sigma_S \tag{11.1}$$

where σ_S represents the total cross section for displacement collisions. At a total neutron dose of 10^{17} (neutrons · volume · time) about 1 atom in 10^6 is displaced in tungsten as a primary event, with a mean energy of 2.5 keV and a maximum energy of 32.5 keV. This primary displacement then induces secondary displacements, and between 30 and 600 displacements are formed per primary. Self-annealing in the displacement spike region might be expected to remove about 50% of the damage.

The amount of damage left is still considerable and readily detectable during field-ion examination. If irradiations to higher total doses are used, problems from overlap of displacement events might be expected to arise.

11.2.3 Field-Ion Examination

Some precautions are necessary in handling irradiated specimens. Usually the activity of the specimens is such that they can be safely handled a few months after irradiation, but at all stages care has to be taken to dispose of electropolishing solutions in a manner which avoids public health risks.

Specimens are electropolished from the irradiated wires by conventional electroetching techniques. In the Cambridge Laboratory a thin layer of electrolyte (aqueous potassium cyanide solution for tungsten; ammonium carbonate solution for iridium) floating on carbon tetrachloride with an a-c potential applied between the specimen wire and a counterelectrode is preferred.[5,6]

Conventional field-ion microscopy is used in the examination with liquid nitrogen as specimen coolant and helium as image gas. A rapid picture sequence is taken between the point at which the image is first seen and the point at which micrographs can easily be taken with a 2-min exposure time.

Using a $4:1$ f 1.0 lens and Tri-X or Scopix G 35-mm film, this exposure time is achieved when the best image voltage is between 10 and 13 kV and the gas pressure is about 3 microns (measured by Pirani gage). At this point the detailed field-ion examination begins. Micrographs are taken so as to have the smallest possible amount of field evaporation between them [in tungsten between 5 and 15 micrographs are taken during the field evaporation of one (110) layer]. Some 500 to 1000 pictures are taken from any one specimen in this manner, and in practice a very high percentage of the atom sites within a volume containing some 10^7 atom sites are imaged. This rather detailed and tedious microscopy has been found necessary to avoid difficulties in interpretation of damage patterns found by field-evaporating specimens one plane at a time rather than a few atoms at a time.

Similar sequences are taken from specimens in the annealed and un-irradiated states to obtain the background imperfection concentration. When a study of the annealing characteristics of the damage is also required, irradiated specimens are given various anneals prior to specimen electro-etching and examination.

Because of the small volume of material studied in a field-ion specimen, it is essential to correlate any observations made on one specimen with further field-ion examinations on other specimens which have been sub-jected to the same irradiation and anneal conditions. It is also desirable to have the best possible image quality so that point defects are positively identifiable. When investigating the annealing characteristics of the damage, it is essential that the image quality from specimen to specimen is kept *constant*.

11.2.4   Interpretation of Field-Ion Patterns

The sequence of field-ion micrographs from any one specimen is examined picture by picture on large prints, and any missing image points are ringed on a first examination. A more detailed examination is then made of the same areas of successive pictures to determine what gave rise to the missing image points. By comparing the pattern of field evaporation from such an area with that seen from the same area of a perfect specimen it is usually possible to identify the source of any irregularities in field evapora-tion behavior.

By using this, as yet qualitative, approach a reproducible count of single vacancies and divacancies, etc., can be made. Analyzing clusters is somewhat more complicated since preferential field evaporation of atoms around the cluster complicates the interpretation. Again, by comparing the picture sequences the two dimensions of the cluster in the surface may be obtained, but the dimension normal to the surface must be essentially derived from the other two.

This approach may appear to be beset with problems and be somewhat hit or miss; however, by applying the method to a number of irradiated specimens statistically useful results are obtained, which compare well with theoretical models of defect distribution in irradiated materials.

11.2.5 Results

Preliminary accounts of investigations of neutron damage in tungsten[7-9] and platinum[10] have been presented. More detailed accounts are either written or being prepared for publication.[11-13]

Two types of study are briefly summarized here:

1. The spatial distribution of vacancies and vacancy agglomerations in the as-irradiated state
2. The annealing characteristics of neutron damage

The as-irradiated state. In tungsten after irradiation to a dose of 10^{17} n.v.t. most of the damage is present as single vacancies and divacancies and give a vacancy concentration of 10^{-4}. There are also a number of vacancy clusters (corresponding to a vacancy concentration in the clusters alone of 10^{-5}); clusters containing around a hundred vacancies, with a nearly spherical shape and no tendency toward collapse, are the most common.[12] These almost certainly represent the remnants of displacement spikes. In one case a 50-vacancy cluster was found some 15 Å from a high-angle grain boundary,[9] which was somewhat surprising in view of the concept of grain boundaries as vacancy sinks.

To date the studies of iridium[11] have been confined to specimens irradiated to very low doses (10^{13} and 10^{14} n.v.t.). It is interesting to note that these low doses correspond to only a few seconds in the reactor, and yet after this short time some of the damage present can be identified with the field-ion microscope. The single vacancy concentration determined on these low-dose irradiation specimens is comparable with the vacancy concentration found in the annealed and unirradiated specimens. Hence to date the study has been directed at determining the nature of the vacancy clusters found.

Three distinct types of large cluster have been observed in the as-irradiated material. The first type of large cluster contains up to about 200 single vacancies in a compact almost spherical form. Large clusters collapse to give dislocation loops. Figure 11.1 shows very clear examples of both these two types of cluster. A section through a compact cluster can be seen at *A*, and a section through a dislocation loop at *B*. The type of contrast observed from the dislocation loop, which is schematically shown in Fig. 11.1*b*, can be qualitatively explained by the stress field of the loop. The distortion of the lattice plane rings is a maximum near the points *A* and *B*

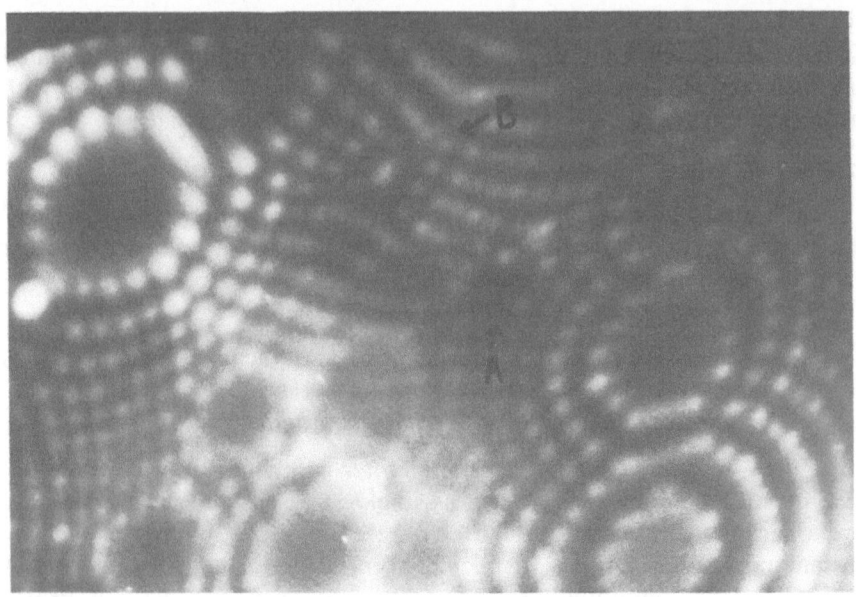

Fig. 11.1(*a*). A compact cluster at *A* is seen very close to a dislocation loop at *B*.[11]

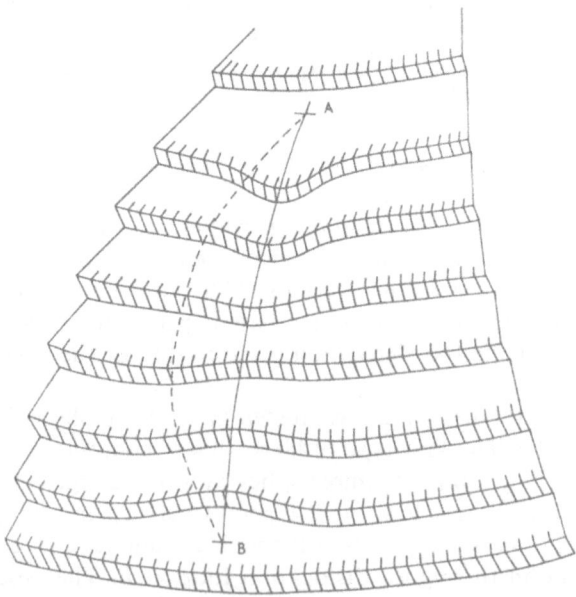

Fig. 11.1(*b*). A schematic diagram illustrating the distortion of the lattice plane rings on the curved specimen surface due to the presence of a dislocation loop intersecting the surface at *A* and *B*.[11]

where the dislocation intersects the surface, and this distortion falls off toward the center of the loop. These loops are found to lie on {111} planes, and it is suggested that they correspond to Frank sessile loops. A determination of the loop size leads to the conclusion that they arise from the collapse of clusters containing about 250 single vacancies.

A third configuration for agglomerated vacancies has also been observed where the single vacancies making up the cluster are dispersed through a small volume of crystal. In one particular case a dispersed cluster was found to contain about 30 to 40% vacant sites in a cylindrical volume 20 Å long having a [100] axis.

It is impossible at this stage to discuss the relative frequency of these types of large cluster. The concentration of primary events after irradiation to these low doses is very small, and the chance of finding such an event in the volume examined by field-ion microscopy is small. Further studies are proceeding on specimens irradiated to higher doses, and thus it should be possible soon to comment on the relative frequency of the three forms of large cluster.

The annealing characteristics of neutron damage. A comprehensive investigation of annealing stages 3 and 4 in reactor-irradiated tungsten has been undertaken.[12,13] The size and spatial distribution of the damage is determined after annealing over a temperature range from reactor ambient to 700°C for various times. The information is obtained in the manner described in Section 11.2.1, and Fig. 11.2 shows an example of how the size distribution is plotted.[12] The size of units used is related to the mean number of vacancies in the cluster in the following manner:

Size range	Average number of vacancies
0.0–0.2	1
0.2–0.4	3
0.4–0.6	10
0.6–0.8	30
0.8–1.0	100

The field-ion evidence is that the total concentration of vacancies remains largely constant during stage 3 annealing but that single vacancies are migrating to give a relatively larger number of divacancies and clusters.[12,13] Further work on plotting the spatial distribution of the damage and the changes occurring during stage 4 annealing is being performed.

11.3 Irradiation Studies on Prepared Specimens

In this section field-ion experiments in which the specimen is prepared prior to irradiation will be discussed. A large percentage of these studies involve irradiation *in situ* in the microscope. A significant advantage of *in*

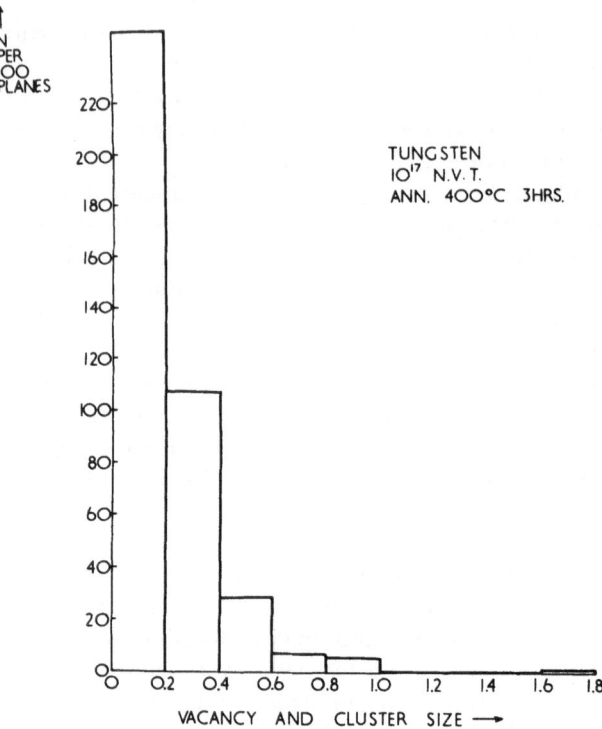

Fig. 11.2. Spectrum of number of events *versus* size for irradiated tungsten. See text for explanation of units.[12]

situ experiments is that the specimen may be held at a predetermined low temperature and the effects of recovery can be minimized.[14] A further advantage is that the irradiation may be conducted (provided the irradiating particles are uncharged) with or without an applied field on the tip. An obvious advantage of having the field applied during bombardment is that the image may be viewed continuously during the irradiation and photographed either in a cine mode, provided some signal amplifying device is used,[14] or periodically when the period between exposures can be correlated with the time between damage events.[15] Surface contamination effects can be minimized in such experiments either by operating the microscope in an ultrahigh vacuum (UHV) mode or by leaving the field on the tip. A disadvantage associated with the latter approach is that the high field stress associated with the imaging field will alter the pattern of damage observed in a way which is impossible to specify.

The following sections describe specific experiments in this field with an emphasis on the experimental techniques used and results obtained.

11.3.1   Low-Energy Neutral Bombardment

Brandon and co-workers have used a low-energy atom source for bombarding specimens in the microscope.[7] The source used was a modified ion gun with a charge-exchange chamber from which argon atoms, having energies in the range of about 100 eV, could be extracted. This source was used in some preliminary experiments aimed at measuring threshold energies for transmitting damage through the specimen. These experiments were performed in an ordinary high-vacuum microscope, and hence the irradiations were performed with the field applied to the specimen.

11.3.2   Low-Energy Ion Bombardment

Some preliminary studies of Xe^+ sputtering of clean tungsten have been made.[16] An UHV field-ion microscope was used and the bombardment performed (in the 100- to 1300-eV range) without a field applied to the specimens. Despite considerable care with the vacuum technique some problems with contamination did occur. However these were sufficiently slight for the damage due to the ion bombardment to be identified.

11.3.3   High-Energy Atom Bombardment

Sinha and Müller built a Penning discharge tube on to the sidearm of a field-ion microscope in order to observe in situ bombardment of tungsten specimens with 20-keV helium atoms and mercury atoms.[17] A Faraday cage was used to monitor the ion current and simple relationships used to derive the number of neutral atoms in the beam. Single vacancies, interstitials, and small clusters were observed. Transmission of damage through the tip was also observed. Stress-enhanced mobility of subsurface interstitials was detected at 21°K, and rapid thermal diffusion from the depth occurred at 90°K.

These experiments were performed in an ordinary high-vacuum field-ion microscope with the field maintained on the specimen during observation. Mass analysis of the discharge beam was used to avoid contamination atoms in the beam.

Good agreement between the experimentally observed damage and the theoretically calculated damage was obtained in this series of elegant experiments. Further experiments of a similar type are underway with the Nelson and Thompson accelerator as a coupled source at Harwell.[18] Another approach is being used in which previously prepared and examined specimens are irradiated in the accelerator and then reinserted into the field-ion microscope for examination of the damage pattern.[19]

11.3.4   Cathode Sputtering

One of the simplest *in situ* radiation-damage experiments is achieved by reversing the potential on the high-voltage generator. As pointed out by Müller, the technique of cathode sputtering has been used for some years to sharpen field emitters by the removal of surface layers.[14,20] In a helium-ion microscope it is possible first to clean the tip by field evaporation and then briefly reverse the potential. During this interval the tip becomes a field-electron emitter at a low field strength of say 30×10^6 V/cm. Helium ions produced by electron–gas collisions near the tip will be accelerated to the tip and strike it with various energies depending upon where they have been ionized. The impact direction will be nearly normal to the specimen surface in all such cases. Müller estimated that at a gas pressure of 2μ something like 60 ions/sec impinge upon each atom of the metal surface. The observed damage was found to compare well with known cathode sputtering yields. Neon or other heavier gas ions would, of course, yield more damage through more efficient energy transfer.

Unfortunately it is difficult, if not impossible, to control the energy distribution and direction of the incident particles by this method. Also the interpretational problems are complicated by complex networks of dislocations which Müller found just beneath the surface.[14]

Despite these disadvantages the simplicity of the experimental arrangement makes further experiments of this type very attractive.

11.3.5   Alpha-Particle Bombardment

Müller[14] more or less concurrently with Brandon and Wald[15] reported studies of *in situ* irradiations with α particles. In both studies 0.5 to 1.0 mCe polonium α-particle sources (5.4 MeV) were positioned to permit irradiation of the specimen from the side. Atoms ejected from the surface and atoms pushed to the surface of the specimen were then detected after analyzing suitably spaced photographic sequences. Interpretations of the patterns obtained are given in the original references based largely on focused collision chains. A reanalysis of the results of the Brandon and Wald experiment concerning the calculated distances over which focused collision sequences might propagate lead to a suggestion that channeling was primarily responsible for the observed patterns.[21] Further experiments of this type in UHV microscopes would remove the interpretational difficulties associated with the presence of the field in these earlier experiments.

11.3.6   Fission-Fragment Irradiation

A study has been made of the track damage caused by fission fragments in metals with the field-ion microscope.[12,21,22] For these experiments irradiation of previously prepared field-ion specimens was used, the irradiation being performed in an evacuated capsule. A small piece of enriched uranium foil faces the specimen inside the capsule, and the geometry is adjusted to produce a beam of fission fragments collimated to better than 7°. The capsules are inserted into an empty fuel element channel in the reactor, and the time of irradiation chosen so as to give an average of one fission-fragment-particle impact per specimen within the field of view. Thin films of molybdenum trioxide are mounted near the field-ion specimens and used to monitor the dose of fission fragments.

Two distinct patterns of damage have been found, depending on the geometry of the irradiation. Where the specimens are bombarded axially, a large (about 250-Å radius) crater is formed. Field evaporation allows the subsurface damage to be investigated and a number of vacancy clusters of about 35 Å diameter have been found.[21] Where the path of the fission fragment makes a high angle with the tip axis, a different form of damage pattern is seen. Very little surface damage is apparent, but below the surface a *vacancy pipe* is found along the particle track which eventually emerges from the side of the specimen.[22]

A considerable extension of this type of experiment is possible and would be extremely useful. This type of experiment naturally allows correlation with electron-microscopy experiments, and the amount of information gained in unit time from such experiments compared most favorably with other field-ion studies of irradiation damage.

11.4   Summary

While a considerable number of exploratory radiation-damage studies have been performed with the field-ion microscope, the number of detailed studies is small and the amount of new information gained is so far not large. These experiments tend to be time consuming and involve considerable interpretational problems. However the field-ion microscope is uniquely suited to damage studies, and it is being applied in a more or less routine way by an increasing number of researchers.

Acknowledgments

The work of the Cambridge field-ion group is supported by the United Kingdom Atomic Energy Authority, Harwell, and the Science Research Council (U.K.) to whom the author is grateful for encouragement and permission to present this account. Professor A. H. Cottrell, Fellow of the Royal Society, has been most generous with helpful discussion and encouragement. The author is particularly grateful to Dr. K. M. Bowkett and Mr. M. A. Fortes for permission to quote freely from their work.

References

1. O. S. Billington, "Radiation Damage in Solids," in: *Proc. Intern. School of Physics*, 1960, Academic Press (New York), 1962.
2. R. Strumans, J. Nihoul, R. Gevers, and S. Amerlinse, "The Interaction of Radiation with Solids" in: *Proc. Intern. Summer School Solid State Physics*, 1963, North-Holland (Amsterdam), 1964.
3. H. G. Van Bueren, *Imperfections in Crystals*, North-Holland (Amsterdam), 1960.
4. Lewis T. Chadderton, *Radiation Damage in Crystals*, Methuen (London), 1965.
5. Y. Yashiro, *8th Field-Emission Symposium*, Williamstown, Mass., 1961.
6. B. Ralph, Ph.D. Thesis, Cambridge University, 1964.
7. D. G. Brandon, M. Wald, M. J. Southon, and B. Ralph, *J. Phys. Soc. Japan* **18** *Suppl.* II: 324 (1963).
8. M. Wald, Ph.D. Thesis, Cambridge University, 1963.
9. K. M. Bowkett, J. Hren, and B. Ralph, *5th European Conf. Electron Microscopy, Prague*, 1964.
10. M. J. Attardo and J. M. Galligan, *Phys. Rev. Letters* **14**: 671 (1965).
11. M. A. Fortes and B. Ralph, *Phil. Mag.* **14**: 189 (1966).
12. K. M. Bowkett, Ph.D. Thesis, Cambridge University, 1966.
13. K. M. Bowkett and B. Ralph, in preparation.
14. E. W. Müller, "Reactivity of Solids," in: *Proc. 4th Intern. Symp. Reactivity Solids*, 1960, Elsevier Publishing Co. (Amsterdam), 1960.
15. D. G. Brandon and M. Wald, *Phil. Mag.* **6**: 1035 (1961).
16. R. W. Strayer, E. C. Cooper and L. W. Swanson, *12th Field-Emission Symposium*, Pennsylvania State University, University Park, Pa., 1965.
17. M. K. Sinha and E. W. Müller, *J. Appl. Phys.* **35**: 1256–1261 (1964).
18. J. A. Hudson, R. S. Nelson, and B. Ralph, experiments in progress at UKAEA, Harwell.
19. J. T. Buswell, experiments in progress at Central Electricity Generating Board, Berkeley.
20. E. W. Müller, in: *Direct Observations of Imperfections in Crystals*, J. B. Newkirk and J. W. Wernick, eds., Wiley (Interscience) (New York), 1962, p. 77.
21. K. M. Bowkett, L. T. Chadderton, H. Norden, and B. Ralph, *Phil. Mag.* **11**: 651 (1965).
22. K. M. Bowkett, L. T. Chadderton, H. Norden, and B. Ralph, *Phil. Mag.* **15**: 415 (1967).

Chapter 12

FIELD-ION MICROSCOPY OF WHISKERS AND THIN FILMS AND APPLICATIONS (REAL AND IMAGINED) TO MASS SPECTROMETRY AND BIOLOGICAL MOLECULE IMAGING†

Allan J. Melmed‡

12.1 Whiskers

12.1.1 Introduction

A wide variety of solids is known to grow under various special conditions in the form of filamentary structures. These structures are called *whiskers*, although the use of the term generally implies that the structure additionally is, or is thought to be, a single crystal with a high degree of perfection. The history of whiskers goes back in time to at least the sixteenth century and includes elemental, alloy, molecular, and polymeric substances. Only in recent years, however, has substantial research interest been aroused. Whiskers may conveniently be classified according to their mode of growth as proper, vapor-grown, chemical-decomposition, melt-grown, solution-grown, mechanical, electrolytically grown, and others. Field-ion microscope investigations have thus far only dealt with chemical-decomposition whiskers and vapor-grown whiskers, and therefore only these will be discussed further at this time. A review of the entire subject has been given by Nabarro and Jackson.[1]

Usually the production of metal whiskers by chemical decomposition is accomplished by the reduction of a vapor-phase halide containing the particular metal. The whiskers grow from nucleation sites on the surface of a generally contaminated substrate. The exact mechanism of growth has not been confirmed, although several theories have been proposed. The main problem is to explain the essentially one-dimensional mode of growth. If

† Contribution of the National Bureau of Standards not subject to copyright.

‡ Institute for Materials Research, Metallurgy Division, National Bureau of Standards, Washington, D.C.

supersaturations sufficiently high to account for the observed growth rates are postulated, it is then necessary to assume that somehow, perhaps through impurity adsorption, the whisker side surfaces (but not the tip) are immunized against nucleation. Alternatively, if the supersaturation is low enough to make nucleation rates very low, then some sort of very efficient permanent sink or sinks must be postulated to exist at the whisker tip so that atoms which diffuse to the tip along the sides become incorporated there. The latter scheme is the more popular one, with emergent screw dislocations usually postulated as the sinks. It should be noted that generally this type of whisker growth is performed under considerably less than ultraclean conditions, regarding vacuum and impurity levels in the starting chemicals and the growth vessels. The whiskers produced, however, have exhibited remarkably high tensile strength and elasticity, and so there is good reason to suspect that they have a high degree of perfection and low dislocation density.

In the controlled production of whiskers from pure vapor, a low supersaturation is maintained by means of a temperature gradient. The whiskers nucleate, grow (increase in length) as an exponential function of time, and then begin to grow linearly and thicken with time. Various theories have been presented, but the scheme which seems to fit the observations best is the one first presented by Sears.[2] This postulates the existence of axial screw dislocations emerging at the whisker tips to provide nonvanishing sinks for atoms diffusing to the tip from the whisker sides. Gomer[3] and Dittmar and Neumann[4] later extended this to a quantitative theory including criteria for cessation of the exponential growth stage and the onset of thickening. Vapor-grown whiskers have been produced under ultrahigh vacuum conditions. However it should be noted that in the growth of vapor-grown whiskers and probably also in whiskers grown by chemical decomposition a rapid initial growth period with the whisker radius constant is followed by a slow thickening stage. During the rapid-growth period the whisker sides may adsorb considerable amounts of contamination compared with the inner core, and during the thickening stage further contamination may occur. Therefore, generally the thinner whiskers of a given batch should be more pure than thicker ones, and the thick whiskers may have an impurity concentration increasing from the central axis outward.

12.1.2 Field-Ion Microscope Studies

Field-ion microscope (FIM) studies of whiskers have thus far only been reported for Fe whiskers[5] grown by chemical decomposition and for Pt whiskers[6] grown from pure vapor. The Fe whiskers were large enough for reasonably direct mechanical mounting but could only be imaged well with the use of neon, or helium with an image intensifier, owing to the low field-evaporation field strength of Fe. The Pt whiskers could be imaged with He

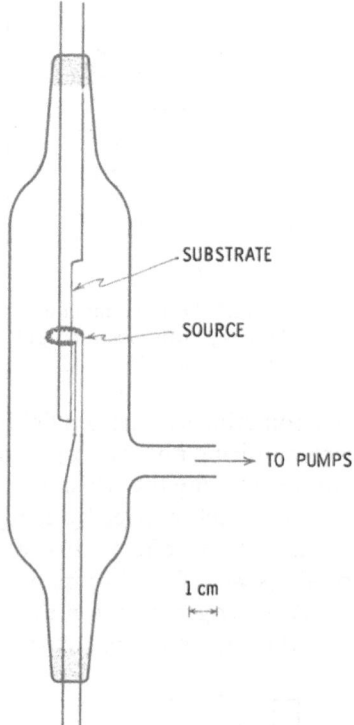

SUBSTRATE

SOURCE

TO PUMPS

1 cm

Fig. 12.1. Schematic diagram of whisker-growth
vessel.

under stable viewing conditions but presented some difficulty regarding
specimen mounting, owing to their small size. The specimen-mounting procedure will be discussed now since it is the only pertinent experimental
technique which is in any way different from the usual ones employed in
field-ion microscopy.

The Pt whiskers were first grown in a simple apparatus, schematically
shown in Fig. 12.1, with the Pt source at about 1600°C and the W substrate
at about 850°C. An all-Pt tip assembly was made by spotwelding a piece of
0.003-in.-diameter Pt wire to a loop of 0.010-in.-diameter Pt wire and etching the smaller diameter wire to a point. This assembly was then clamped to
the moving stage of an optical microscope and heated to about 800°C by
passing an alternating current through it. The whisker growth substrate was
then moved with a pneumatically operated micromanipulator to the position
shown in Fig. 12.2a. Next the Pt tip was brought into contact with a whisker
and retracted, as shown in Fig. 12.2b and 12.2c, and then heated to about
1000°C to strengthen the whisker-tip weld. The strength of this weld was
finally tested by putting it into a continuous stream of tap water. Figure 12.3
illustrates the mounting technique for observation of the whisker-growth

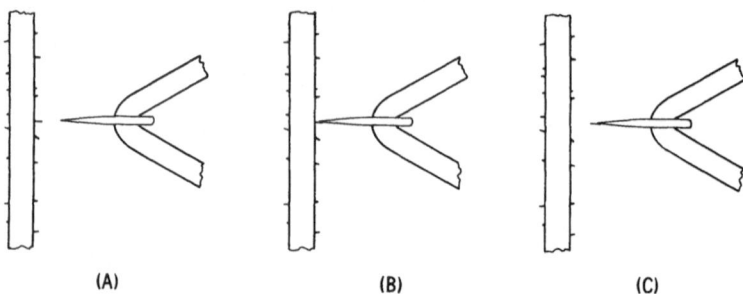

Fig. 12.2. Whisker mounting technique: (a) initial positions of Pt point and whisker-growth substrate, (b) Pt point contacting a whisker, and (c) Pt point and whisker retracted.

end. Sometimes hydrogen field-ion microscopy could be done with the whiskers' requiring no further preparation. However helium field-ion microscopy required etching the whiskers to obtain sharper points. This was done under the optical microscope by using the microscope stage and micromanipulator to achieve the effect illustrated in Fig. 12.4. (In later experiments KCl solutions were found to be preferable to KCN solutions.)

A hydrogen field-ion micrograph of a Pt whisker is shown in Fig. 12.5. An {011} plane is approximately in the center, the large dark regions left

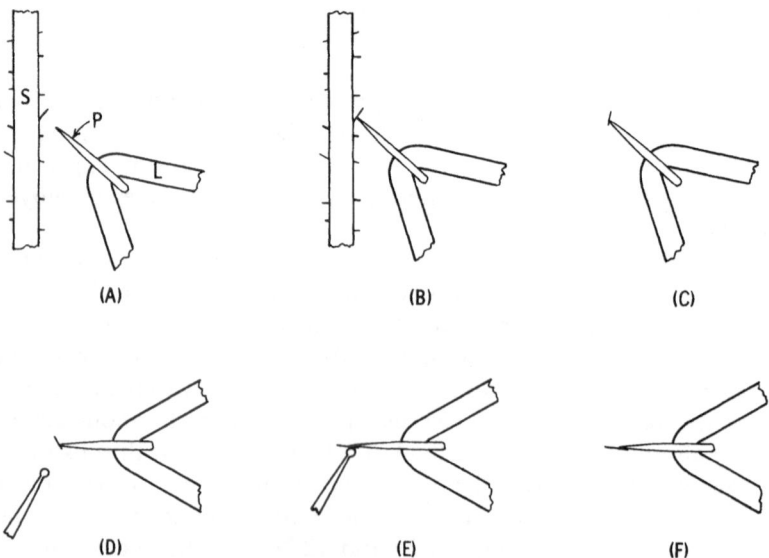

Fig. 12.3. Mounting technique for observation of growth end of whisker: (a) Pt point P on Pt loop L positioned near whisker-growth substrate S, (b) whisker contacted, (c) whisker removed, and (d, e, f,) optional alignment procedure.

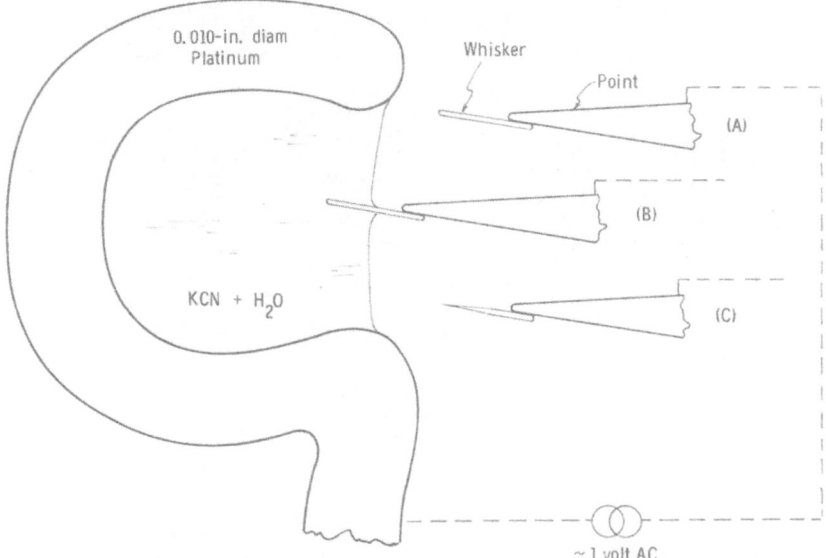

Fig. 12.4. Whisker sharpening technique.

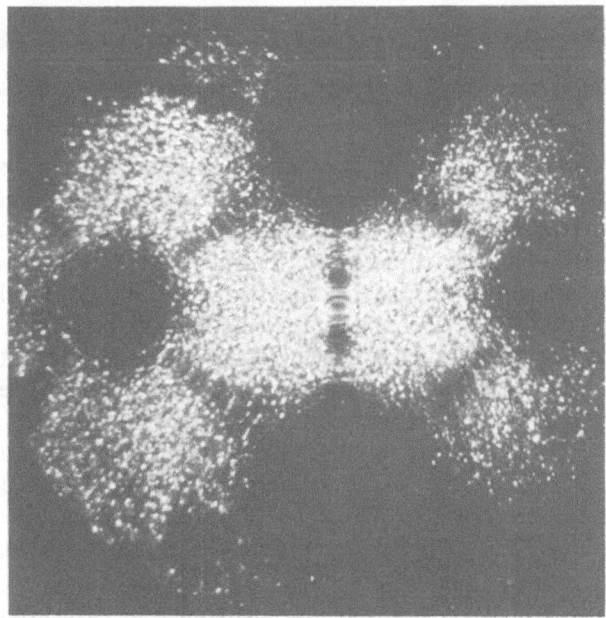

Fig. 12.5. Hydrogen field-ion micrograph (77°K) of thermally smoothed Pt whisker; {011} plane near center.

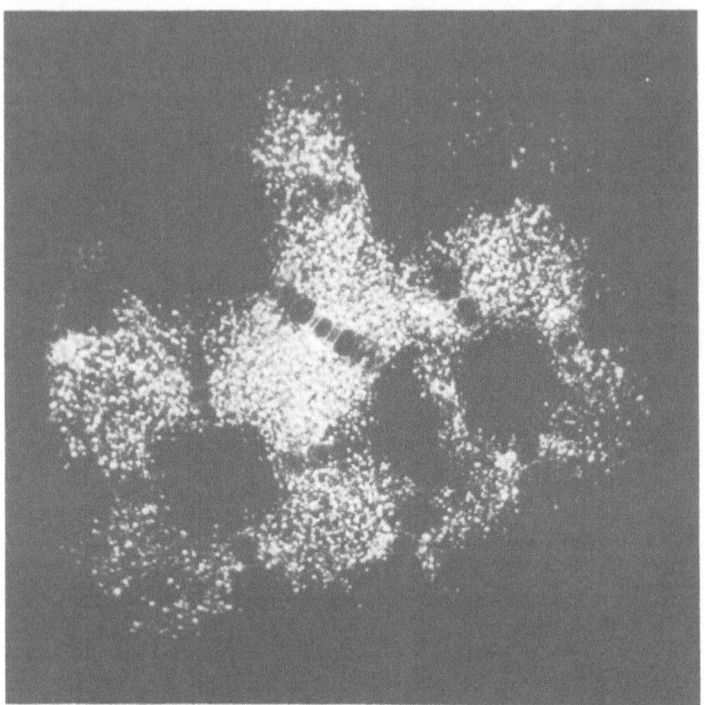

Fig. 12.6. Hydrogen field-ion micrograph (77°K) of a twinned Pt whisker.

and right of center are {001} regions, and the larger dark regions above and below center are {111} planes. The enormous extent of disarray of the surface atoms is due to the fact that this emitter was thermally cleaned and smoothed (in vacuum). The elevated temperature treatment (about 1200°C in this case) causes a large amount of disorder in the surface, and much of this disorder remains after cooling the specimen. Only the smooth low-index planes and a few zone lines retain geometric regularity on an atomic scale. Under these conditions a search for crystal imperfections is certainly impeded over a large part of the surface. However much larger fractions of whisker-tip surfaces could be studied in this manner compared with helium field-ion microscopy, for which the maximum tip radii are about half that for hydrogen-ion microscopy. The tip radius in Fig. 12.5 is about 2.8×10^{-5} cm. Figures 12.6 and 12.7 show hydrogen field-ion micrographs of a twinned whisker and whiskers with indications of possible emergent screw dislocations on central {011} planes. The occurrence of twin boundaries was found to be quite common in vapor-grown Pt and Au whiskers studied[7] by field-electron and ion-emission microscopy, twins appearing in some

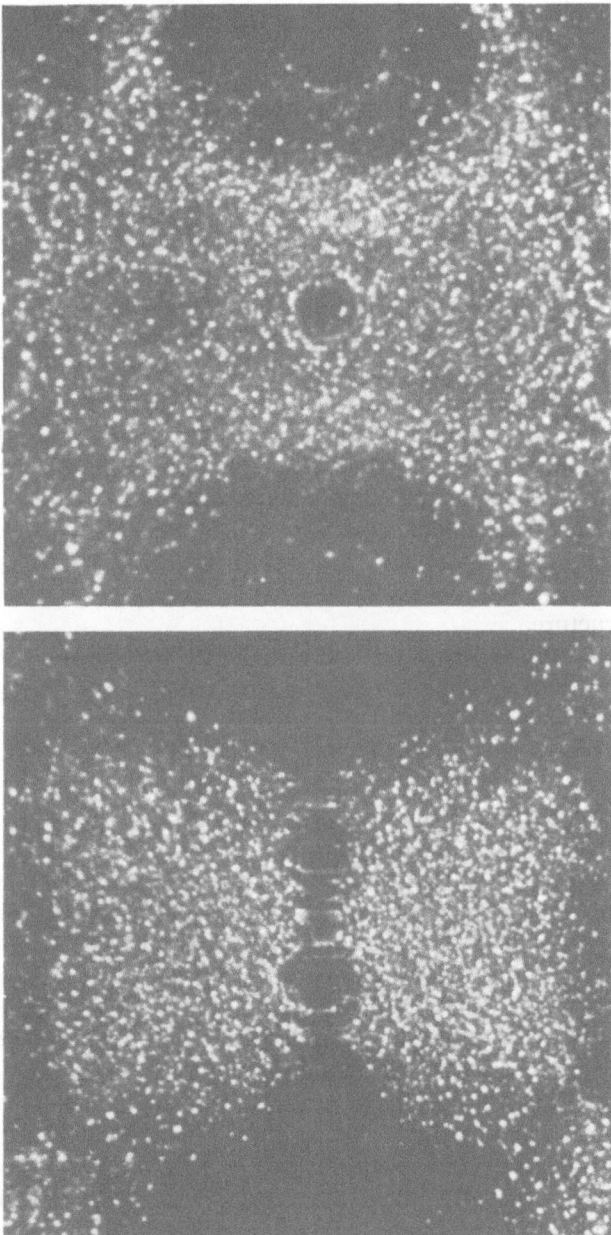

Fig. 12.7. Hydrogen field-ion micrographs (77°K) of Pt whiskers with imperfections on central {011}, possibly due to emergent screw dislocations.

30% of the whiskers studied. Evidence of the presence of screw dislocations was found in three of the eleven Pt whiskers studied by hydrogen FIM. Further analysis of the dislocations observed by hydrogen field-ion microscopy could not be done since the emitters were too blunt to allow field evaporation to be performed.

Whiskers which were sharp enough for helium field-ion microscopy were so examined at liquid-nitrogen temperatures. Field evaporation could be performed, and 13 whiskers were observed through several hundred atom layers each. In no instance was a central screw dislocation observed. As seen in Fig. 12.8, which shows a helium field-ion micrograph, image intensity was very low over a large part of the image, and it is possible that dislocations in these regions could have gone unnoticed.

The FIM study of Fe whiskers[5] done by E. W. Müller and O. Nishikawa found some screw dislocations and line imperfections also. The authors were able to come to qualitative conclusions about the relative degree of perfection of whiskers grown under different conditions by comparing the degree of perfection achievable by field evaporation. However it was not possible to draw quantitative conclusions about defect density and impurity concentration, owing in part to the generally poor development of surface structure.

The recent improvement in field-evaporated surface structure of some metals brought about by the introduction[8] of He–H_2 imaging–gas mixtures should make possible improved studies of Fe and other low-melting metal whiskers. (See also Chapter 6.) However there remains one large obstacle to quantitative studies with whiskers: Generally the whisker must be sharpened. If one considers 1.5×10^{-5} cm and 3×10^{-5} cm as practical upper limits of the tip radii for helium-ion microscopy and hydrogen-ion microscopy respectively (use of most other imaging gases such as neon and nitrogen would be intermediate cases), the maximum tip areas imaged allow only small fractions of the total whisker cross section to be examined. For example, a very thin chemical decomposition whisker, having a diameter of 2×10^{-4} cm, would have only about 2% of its cross-sectional area exposed for study in a helium FIM, and about 9% exposed to a hydrogen FIM. The generally smaller whiskers grown by pure vapor deposition have allowed larger fractions of the cross-sectional area to be observed but still only a few percent by the more analytical helium FIM. Furthermore, as illustrated in the electron micrographs of Fig. 12.9, the point resulting from the tip-etching process is generally off center. This means that information obtained from such specimens regarding defect density, impurity concentration, and presence of axial screw dislocations cannot be considered necessarily of statistical significance. The problem is probably soluble with some experimental inconvenience. Only whiskers thin enough for FIM

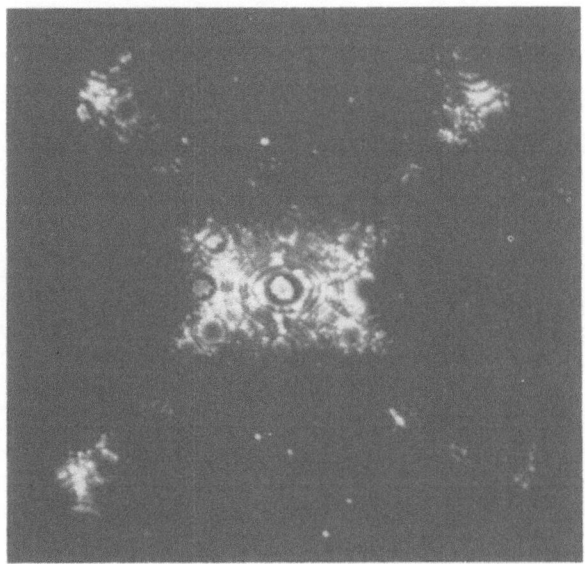

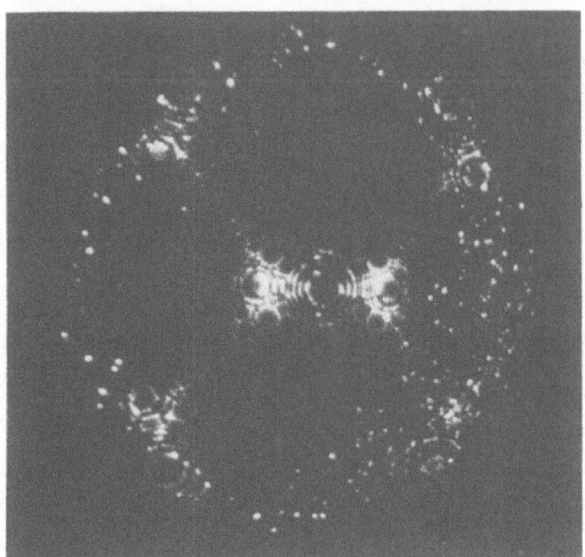

Fig. 12.8. Helium field-ion micrographs of field-evaporated Pt whiskers (solid-nitrogen cooling); {011} planes in centers.

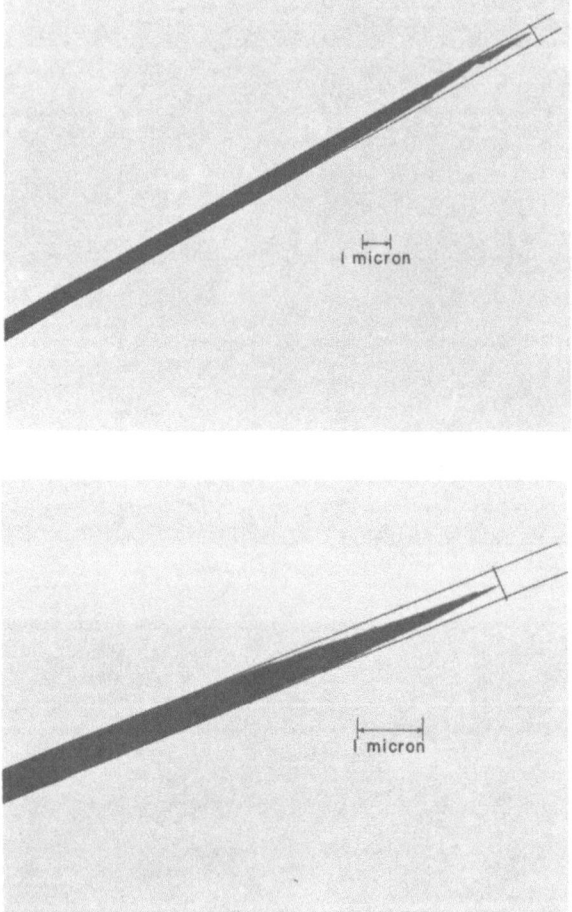

Fig. 12.9. Electron micrographs of sharpened Pt whiskers. Tips
are off center, as seen in reference to lines drawn parallel to
whisker sides.

studies without additional sharpening should be used, and all whiskers that
have been sharpened should be geometrically characterized by optical or
electron microscopy.

12.2   Thin Films

Field-ion microscopy of thin films is an area of study which is just
beginning. In principle it is an ideal subject for investigation by FIM since
it provides specimens which, in many instances, could be totally examined,

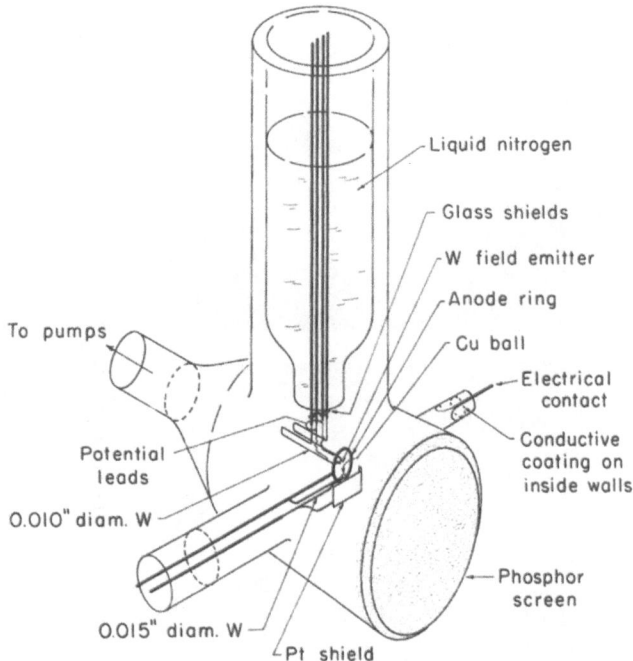

Fig. 12.10. Field-electron emission microscope for studying epitaxial growth on an emitter tip; schematic diagram.

i.e., examined in each layer through their entire volumes. The subject is of considerable practical as well as academic interest because of the increasing number of thin-film devices being developed in the electronics industry. For the most part the films are prepared by epitaxial growth techniques.

Recently a technique was demonstrated[9] for growing thin films epitaxed from metallic vapor on the conventional emitter tip of a field-electron emission microscope. It has thus far been successfully used to grow single crystals and polycrystals of Ag, Cu, Pb, Y, and Fe on W point substrates. Figure 12.10 schematically depicts a microscope used for these experiments. Minor modifications, such as the addition of a Cu cooling mantle and a change from a two-lead to a four-lead evaporator unit (for better degassing), convert this design to a combination field-electron and field-ion microscope for ultraclean experiments. Nucleation and epitaxial crystal growth experiments have been carried out for the most part on thermally cleaned and smoothed surfaces by field-electron emission microscopy. Figure 12.11 shows field-electron micrographs of Cu nuclei, an Fe crystallite, and a completed Y single crystal, all on W substrates.

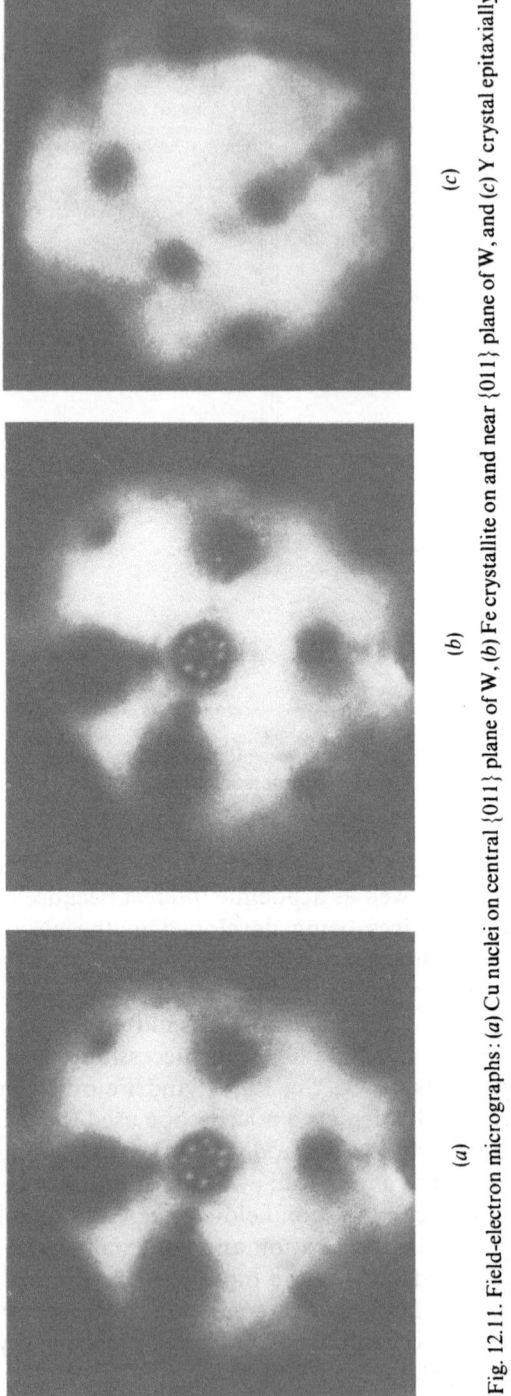

(a)

(b)

(c)

Fig. 12.11. Field-electron micrographs: (a) Cu nuclei on central {011} plane of W, (b) Fe crystallite on and near {011} plane of W, and (c) Y crystal epitaxially grown on W.

Just a few FIM studies have been made thus far which relate to the growth and properties of thin films; some thickness measurements of epitaxial Cu films have been made,[9] and certain aspects of the initial nucleation process have been approached. E. W. Müller[10] observed W atoms condensed on W; T. Gurney, Jr., F. Hutchinson, and R. D. Young studied[11] the same system, and T. Rhodin, H. Montague-Pollock, and M. Southon have reported[12] preliminary observations of Au deposition on W and Fe deposition on Ir. Figure 12.12 shows 21°K-helium field-ion micrographs of W condensed on W from the work of Gurney, Hutchinson, and Young. At 3000°C W was evaporated from a source positioned to the right of the emitter, as oriented in the micrograph. The initial deposit was then carefully field evaporated, during which a series of micrographs were made. Analysis of the micrographs led to the conclusion that condensation of W on a W tip at 21°K (or 77°K) from a source approximately at a right angle to it produced a deposit whose first layer was only two-thirds to three-fourths complete and developed into discrete nuclei as further layers were added. From a comparison of the experimental results with Monte Carlo and statistical calculations, R. D. Young and D. C. Schubert further concluded that the W atoms which impinged upon the W tip remained in the first potential wells encountered.[13]

A large range of substrate temperatures remains to be studied, as well as a range of other materials, from the point of view of gaining information about initial nucleation events. Potential areas of investigation where FIM techniques could provide useful information are:

1. Structure of nuclei in the 2- to 10-atom size range, as a function of substrate atomic geometry
2. Structure and purity of the crystal–substrate interface layer(s), particularly when growth is epitaxial for material A on substrate B
3. Defect density and impurity levels for epitaxed crystals
4. Nature of the interface between epitaxed crystals after coalescence

As usual it is to be anticipated that interpretative problems will have to be overcome, particularly for item 1 above.

12.3   Field-Ionization Mass Spectrometry

In the conventional mass spectrometer the material to be analyzed is ionized by electron impact in a small region of the spectrometer. Electrons are produced by thermionic emission and normally are accelerated through about 70 V. Collisions of electrons with molecules cause the formation of (mostly) positive ions, which are then accelerated into the analyzer section of the spectrometer. Line-width broadening of the mass peaks is caused by the spatial distribution of ionization events inherent in the functioning

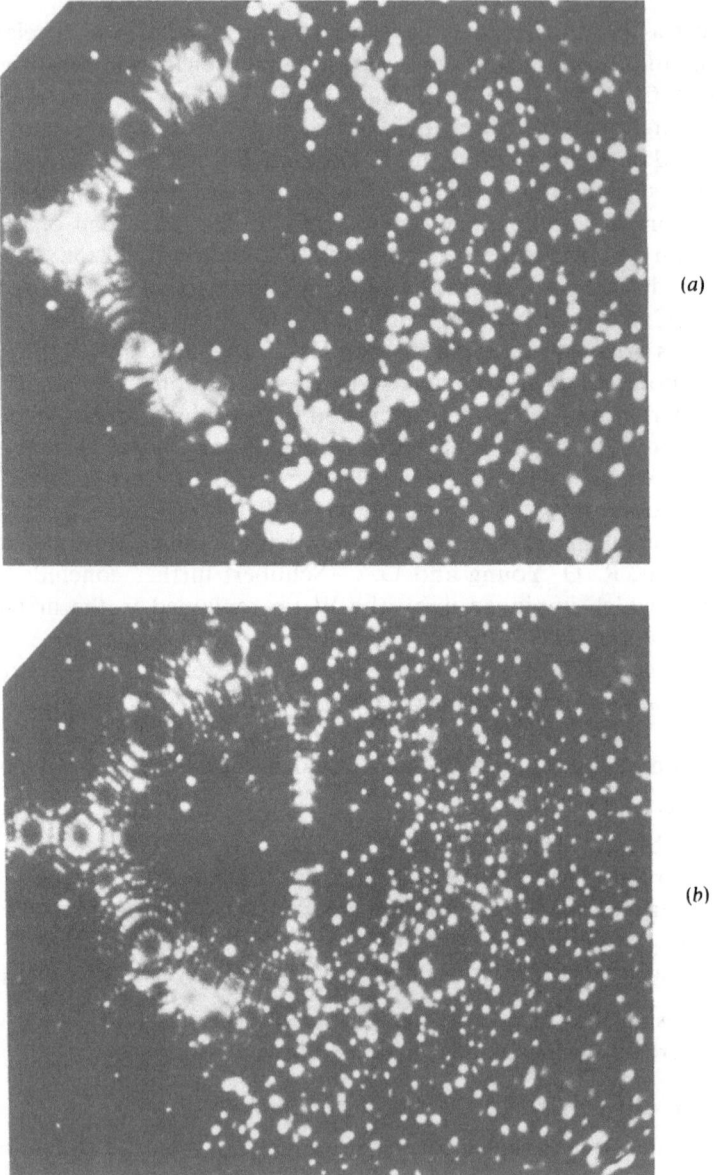

(a)

(b)

Fig. 12.12. Helium field-ion micrographs (21°K) of W condensed on
W: (a) condensed W averaging about three atoms thick on right side
of tip, and (b) after removal of some of the outermost condensed W
atoms by field evaporation. (Courtesy R. D. Young, U.S. National
Bureau of Standards.)

of electron impact sources, but more troublesome is the production of fragment ions resulting from electron impact dissociation. This fragmentation can be reduced by operation of the electron source at lower electron energies, but lowering the electron energies too much (close to the molecular ionization energies) introduces ion-current instability.

12.3.1 Field-Ion Source

The first field-ion source used for mass spectrometry, introduced by Inghram and Gomer,[14] consisted of a conventional FIM with a hole in the screen, which allowed some of the ions formed at or near the tip to enter the analyzer section of the mass spectrometer. This arrangement provides directional selectivity among the ions formed at or near various tip regions but severely limits the magnitude of the ion current entering the analyzer. As first suggested by Gomer and Inghram,[15] Müller and Bahadur[16] developed a focusing system so that a much larger percentage of the ionized particles could be analyzed. Later instrumental developments consisted mainly of improvements in the optics and the introduction of wire field-ion emission sources[17] but were unsuccessful because of abundant surface irregularities and protrusions (possibly whiskers).

In discussing the properties of a field-ionization mass-spectrometer source it is convenient to start by considering the point type of source. The theory of field ionization was treated in detail in an earlier section. That field ionization is the mechanism of ion production in the FIM was first suggested[14] as a conclusion of the early field-ion mass spectrometer experiments of Inghram and Gomer, in which they observed asymmetric line-width broadening as a function of tip field strength. Such line-shape changes may be interpreted as follows: The relationship of ion mass m, charge e, accelerating voltage V, magnetic field H, and spectrometer path radius R for a magnetic deflection instrument is given by $m/e = (H^2 R^2/2V)$. For constant H, R, and e this reduces to $m = K/V$, with $K = (H^2 R^2 e)/2$. Suppose the tip (ionization source) is kept at voltage V_t at the entrance port of the spectrometer analyzer. Any ions of mass m_0 that form at the tip surface will then be focused and collected when the accelerator voltage is some value depending on the numerical value of K and m_0. Ions of mass m_0 formed at the tip surface gain energy eV_t before going through the entrance port. Ions of mass m_0 that form some distance X away from the tip only gain energy $eV_t - eFX$, where F is the field strength (by assuming $V = FX$), i.e., they are accelerated through a smaller potential drop before reaching the entrance port. Consequently a higher instrument accelerating potential is required to focus these ions, which causes the apparent mass to be less than m_0 and broadens the nominal m_0 peak. The point type of source thus enables the geometrical origin of ion species to be determined. From line shape measurements it was determined[14] that the ions contributing to a hydrogen FIM image predominantly originated from a zone within 5 Å of the tip surface.

Since the ionization is brought about by electron tunneling from the molecules, very little or no molecular fragmentation occurs, and this is the main advantage of using a field-ionization source. For example, Fig. 12.13 compares the mass spectrum of pentane by field-ionization with its spectrum by electron-impact ionization. Some dissociation in the high electric field does occur in certain cases, though, and this will be discussed below.

Problems arise in the use of point type of field ionization sources owing to the very small size of the emitter. Thus, the ion current is small and there usually are large (20% or greater) current fluctuations due to fluctuating ionization conditions. These occur from changes in the local geometry of adsorbed material and from chemical etching effects[18] on the tip and are

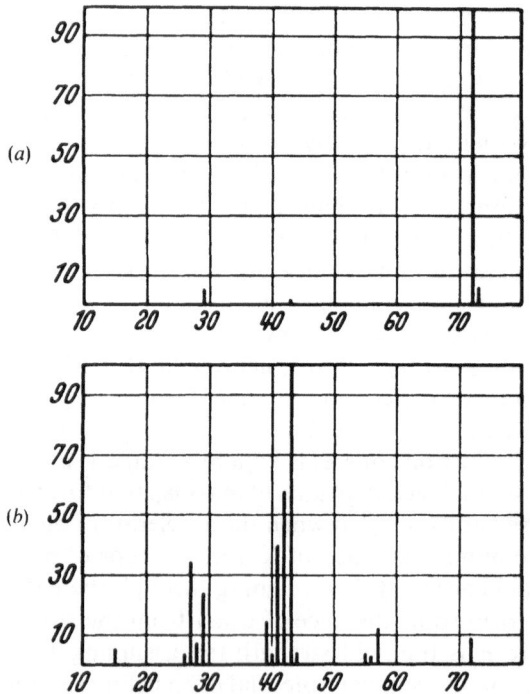

Fig. 12.13. Mass spectra of pentane: (a) field-ionization source; and (b) electron-impact ionization source. (Courtesy H. Okabe, U.S. National Bureau of Standards.)

exaggerated because the small size of the tip does not allow much statistical averaging. Consequently, some advantages are to be had by the use of a cylindrical wire ionization source instead of the point type.

The feasibility of using wires as field-ion emission sources in a mass spectrometer was first demonstrated by H. D. Beckey.[17] He used 2.5×10^{-4}-cm-diameter Pt Wollaston wires, with about a 3 to 5-mm length of exposed Pt, as a compromise between the mechanical weakness of thinner wires and the excessively high voltages required for ionization with thicker, more stable wires, and he also found that quartz fibers covered with evaporated films of Ni or Pt were quite suitable.

The electrostatic field at the surface of a smooth cylindrical wire of radius r_c in vacuum with the other electrode a concentric cylinder of radius R is

$$F = \frac{V_c}{r_c \ln R/r_c} \tag{12.1}$$

where V_c is the potential difference across R and r_c. Compared with the situation for ionization at a point type of emitter where

$$F \cong \frac{V_t}{5r_t} \tag{12.2}$$

$$\frac{V_c}{V_t} = \frac{r_c \ln R/r_c}{5r_t} \tag{12.3}$$

for the same field strength. For $r_c = 1.25 \times 10^{-4}$ cm, $r_t = 5 \times 10^{-5}$ cm, and $R = 0.5$ cm, this would imply that cylinder ionization voltages would have to be about four times greater than tip ionization voltages. However in practice the wires are not smooth perfect cylinders, so that there is local field enhancement at many surface regions, and certain molecules having low ionization potentials can be ionized with reasonably low voltages (10 to 15 kV). This same geometric circumstance, however, causes the actual ion current to be about 100 times less than predicted on the basis of the nominal diameter of the wires. There is still an intensity gain of about 100 compared with that of typical point sources, and there are considerably smaller intensity fluctuations.

Molecular fragmentation sometimes occurs with a field-ionization source. This is influenced by surface adsorption effects as may be seen by comparing spectra of the same substance made with the use of a point type of field-ionization source in one case and a wire source in the other. Generally, more fragmentation occurs when a wire source having typically about 10^4 times greater surface area is used. However fragmentation has also been observed in using point type of sources, and a mechanism for field dissociation has been advanced by Beckey.[19] The mechanism is a general one, but may be understood simply in the field dissociation of an H_2^+ ion. In the presence of the electric field gradient near a tip at positive potential the ion becomes polarized, i.e., there is a greater electron cloud density at the end of the ion near the tip compared with that at the far side. The ion may then be considered to consist of a polarized neutral H atom (near side) and an H^+ ion (far side). The polarized neutral atom is weakly attracted toward the tip, and the ionized atom (proton) interacts with the electric field in a manner similar to the field desorption interaction described by Gomer.[20] The fact that field dissociation occurs after field ionization events in a geometrically small and well-defined region, i.e., in a narrow zone near the tip, has enabled the measurement of lifetimes of short-lived metastable molecules to be extended from about 10^{-6} to 10^{-13} sec.[21] A detailed discussion of this and other significant applications of the field-ionization mass-spectrometry technique is outside the intended scope of this section.

12.4   Biological Molecule Imaging

12.4.1   The Goal

Ever since the early days of field-emission microscopy (FEM) researchers working with the instruments have dreamed of someday developing their capabilities to the extent of seeing individual atoms and molecules with them. Relatively early in FEM history, E. W. Müller demonstrated that "images" of certain organic molecules could be obtained by adsorbing these molecules on the emitter of a field-electron emission microscope (FEEM). Later, after Müller introduced the field-ion microscope, which has been capable of seeing atoms of a metal and occasionally a semiconductor surface, interest began to rise in applying the FIM to the imaging of large organic molecules such as RNA and DNA. Although the ultimate goal may be considered a complete mapping, in atomic detail, of these molecules, the more immediate goal seems to be considerably less demanding. A molecule profile outline in, or nearly in, atomic detail is sought.

12.4.2   Early Experiments

The first study of organic molecules with the FEEM was reported in 1950 by E. W. Müller.[22] An organic dye, Cu–phthalocyanine, consisting of thin planar molecules of about 10 Å in lateral dimensions with fourfold rotational symmetry, produced discrete bright patterns when adsorbed on a W field-electron emitter tip. Many of these patterns were quite similar to the molecular shape expected on the basis of X-ray determinations,[23] appearing as four-leaf-clover (quadruplet) patterns. Some were bright doublets, or single spots, and occasionally a doughnut-shaped pattern, or some odd shape would appear. It was concluded that these patterns resembled individual Cu–phthalocyanine molecules in different positions on the emitter surface. Further investigations[24–28] revealed the puzzling fact that despite the wide variety of molecular structures possessed by the many organic materials (and tungsten oxide) studied, no different image shapes occurred, and it appeared that the molecular patterns could not be interpreted as simple molecular images. Later experiments[29] also demonstrated that the patterns often represented more than one molecule. Several mechanisms have been proposed[24–32] to explain the origin and interpretation of molecular patterns in the FEEM, but a critical review of these is outside the scope of this presentation.

A field emission microscope used to study organic molecules is schematically shown in Fig. 12.14. Some molecular patterns from Cu–

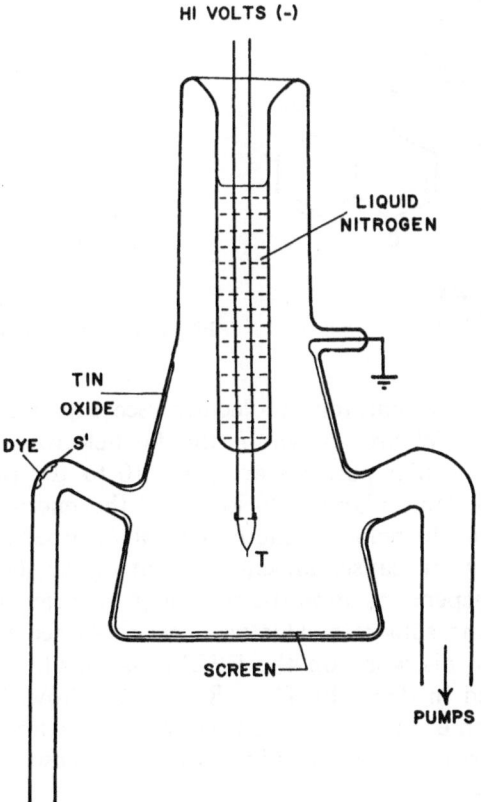

Fig. 12.14. Field-emission microscope for studying organic molecules; schematic diagram.

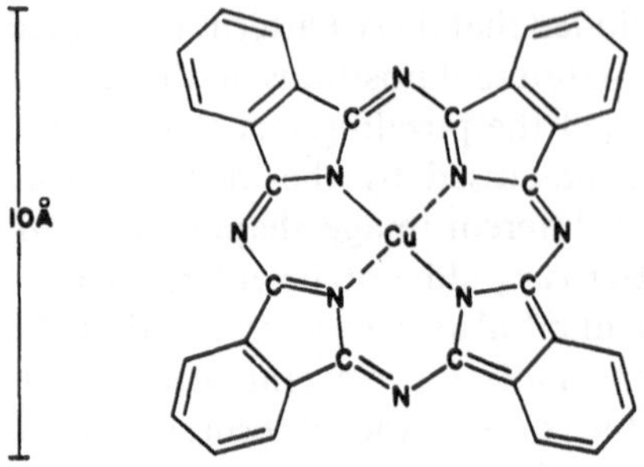

Cu-phtalocyanine

$C_{32}H_{16}N_8Cu$

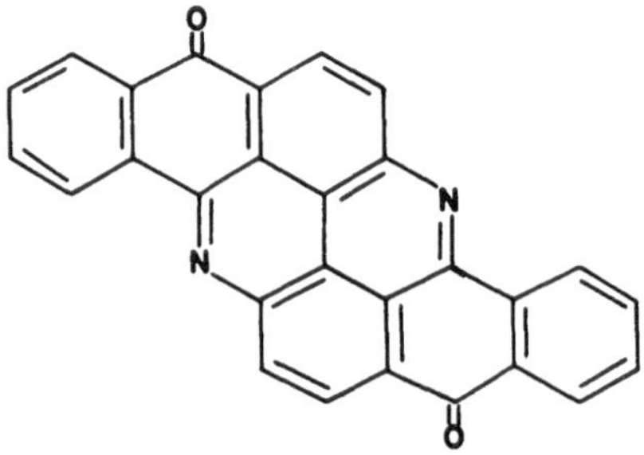

Flavanthrene

$C_{28}H_{12}N_2O_2$

Fig. 12.15. Schematic molecular diagrams.

phthalocyanine and flaventhrene molecules (See Fig. 12.15) adsorbed on thermally-smoothed W are shown in the field-electron micrographs of Fig. 12.16. The molecular patterns are some 10 to 100 times larger than predicted from the molecular dimensions and the microscope magnification. Of relevance to the field-ion microscopy of adsorbed species is the fact that a local protrusion causes an enlargement of the local image by an amount which is dependent upon the relative geometries of the protrusion and the surrounding substrate. This was first explained by E. W. Müller in connection with his work on the FEEM visualization of Ba atoms,[33] and later described in detail by D. J. Rose, who showed[34] that a hemispherical protrusion of radius ρ_0 on a tip of radius R with a normal magnification M_0 caused the local magnification to increase to approximately $M = 1.1(R/\rho_0)^{\frac{1}{4}}M_0$.

To assess the feasibility of imaging organic molecules in the FIM, organic-molecule–field-desorption experiments were performed[28] in the

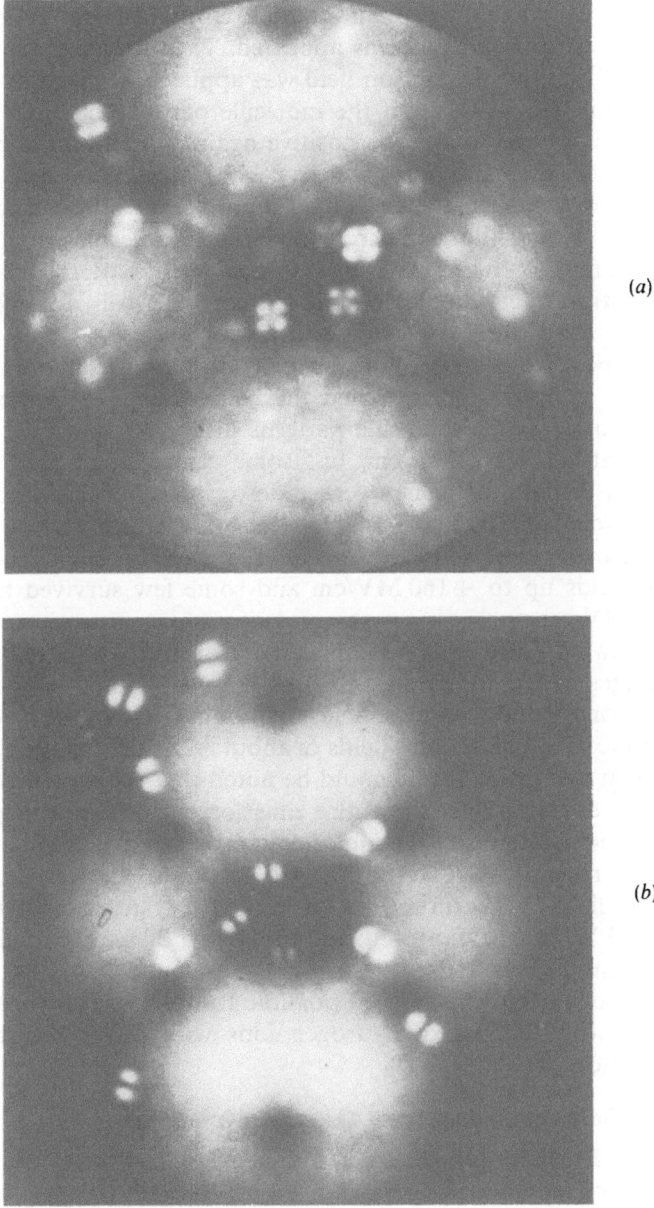

Fig. 12.16. Field-electron micrographs of organic molecules adsorbed on W at 77°K: (a) Cu–phthalocyanine on thermally smoothed W, and (b) flavanthrene on thermally smoothed W.

tube of Fig. 12.14 with the addition of a copper cooling mantle. Cu–phthalocyanine was sublimed onto the field emitter (W and Pt tips were used) until 15 to 20 molecule patterns appeared. Then the voltage was reversed, and a small positive desorption field was applied. Next the tip voltage was made negative again to see if the molecule patterns had survived. If they had, the tip voltage was made positive again, slightly more positive than the previous time. This stepwise process was continued until all of the patterns were gone. Three effects tend to interfere with obtaining a smooth relationship giving the number of remaining molecules as a function of desorption field strength: (1) The small number of patterns involved causes large statistical fluctuations, (2) some patterns normally disappear when the tip voltage is reduced to zero,[30] and (3) some new patterns are made to appear by positive fields of about 80 to 100 MV/cm.[30] The results are therefore stated in broad terms. When thermally annealed W tips were used (at 77°K), most of the molecule patterns disappeared at field strengths of less than about +160 MV/cm, but some remained at field strengths of +190 to +200 MV/cm. The average desorption field for 43 doublets was +154 MV/cm, and the average for 15 quadruplets was +126 MV/cm. When thermally annealed Pt tips were used, most of the molecular patterns survived fields up to +160 MV/cm and some few survived fields as high as +210 MV/cm.

In view of these results it appeared reasonable to attempt to obtain FIM images of Cu–phthalocyanine on thermally smoothed tips by using mercury ions, which require fields of about +70 MV/cm, or by using hydrogen ions, which require fields of about +200 MV/cm. The latter value might seem marginal, but it should be noted that the measured desorption fields were averages over the entire emission region and it is possible that the local field strengths at the molecular species were greater.

More recently R. C. Abbott and W. A. Livingston, Jr., reported[35] desorption field F_d measurements for removing transfer RNA from a field-evaporated W surface. For d-c fields applied for 30 sec they found $F_d = 119 \pm 8$ MV/cm, and for pulsed desorption they found $F_d = 187 \pm 5$ MV/cm. They concluded that it might be possible to observe FIM images of these molecules with the use of hydrogen ions with pulsed fields and image intensification.

12.4.3   Field-Ion Imaging of Organic Molecules

Some FIM images of organic molecules have been obtained,[30] but resolution of atomic details has not been achieved. Imaging with mercury only produced randomly located diffuse bright spots with no discernible details. Hydrogen field-ion microscopy was then pursued, despite the higher imaging field strength required. Comparison could be made of ion images and electron emission images taken sequentially since the hydrogen could be pumped away quickly and small amounts of residual hydrogen would not cause trouble by cathode sputtering during the field-electron emission mode of operation. The microscope used was the one shown in Fig. 12.14 plus a Cu cooling mantle and was always baked out prior to use. Figure 12.17 shows typical hydrogen field-ion micrographs (liquid-nitrogen cooling) of Cu–phthalocyanine adsorbed on thermally smoothed W and Re. Positive correlation was obtained between ion images and field-electron molecule patterns taken before or after, for about 50% of the ion images.

The hydrogen field-ion images of thermally smoothed W and Pt surfaces with adsorbed Cu–phthalocyanine occurred at about 80% of the field strength required for imaging the same surfaces without the adsorbed molecules. The ion images of molecules usually consisted of single spots or doublets, about five to ten times smaller than the corresponding electron-emission molecule patterns. No improvement in the ion images was observed when liquid-hydrogen cooling was used instead of liquid-nitrogen cooling.

Some experiments were also performed[30] with the use of a 60-cycle a-c and d-c-biased power supply.[36] After admitting hydrogen into the microscope, the voltage cycle was adjusted so that a small negative portion produced electron emission and a larger positive part of the cycle produced an ion image. Thus the screen simultaneously displayed the electron-emission image and the hydrogen-ion image of Cu–phthalocyanine on W. The large amount of hydrogen required for the ion imaging made the molecules extremely unstable caused by tip bombardment during the negative part of the voltage cycle. No image intensifier was available at the time, so no photographs were obtained. However, the combined images could be visually observed for 15 to 30-sec periods, and many instances of distinct electron-emission molecule patterns superposed on bright ion image spots were noted.

Nylon sublimation products were examined[37] in the FEEM and the FIM by warming a short cylindrical length of the material in the microscope after achieving high-vacuum conditions (about 1×10^{-8} torr). Figure 12.18 shows field-electron micrographs and a hydrogen field-ion micrograph (liquid-nitrogen cooling) of nylon on thermally smoothed W. The electron-emission patterns were generally unstable with movement and rearrangement occurring among individual parts of a given "molecule" pattern.

12.4.4 Recent Work

Recent work has focused on the possibility of imaging molecules of biological interest. These molecules are quite large (hundreds of angstrom units) on the size scale of typical FIM tips. It is believed that the simple

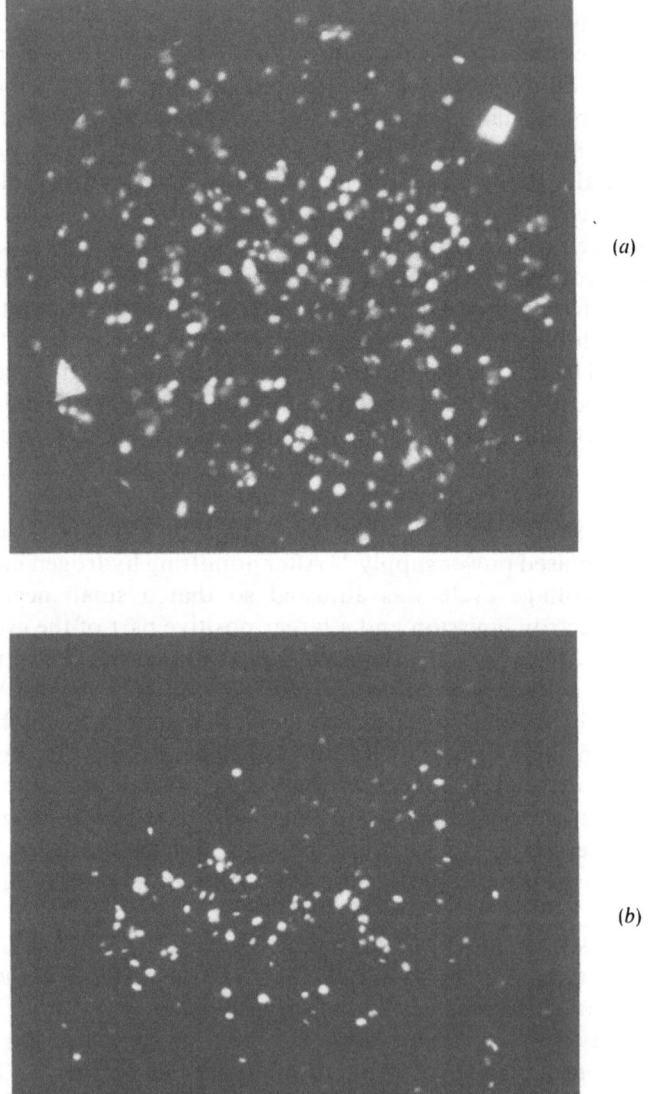

(a)

(b)

Fig. 12.17. Hydrogen field-ion micrographs (77°K) of Cu–phthalocyanine: (a) on thermally smoothed W, and (b) on thermally smoothed Re.

sublimation methods employed in the earlier work on smaller organic molecules will not suffice to give meaningful results since thermal degradation is predicted. Therefore the initial research has been concentrated on ways to put biological molecules on field-ion emitters without damaging the molecules. H. Montague-Pollock has reported[38] a specimen delivery scheme for selectively bringing large nonvolatile molecules to the tip in a gaseous helium carrier. A dilute liquid solution of the molecules to be studied is blown through an atomizer with helium gas, and the droplets so formed evaporate which leaves the large molecules and contaminant moving with the helium stream. This mixture is then passed through a trap cooled with liquid hydrogen. When conditions are properly adjusted, the lighter molecules in the stream are selectively trapped because of differences in their Brownian motion compared with the movement of the large molecules. The stream then passes by the tip, on which the large molecules may be adsorbed.

R. C. Abbott has recently discussed[39] the specimen deposition problem. He cites the following as possibly applicable methods: simple drying, freeze-drying, freeze-drying of sprays, embedding in another matrix, and sublimation of molecule beams. The work of Gurney, Hutchinson, and Young,[11] discussed in an earlier section, was done to aid in an evaluation of an embedding technique, i.e., embedding, or shadowing the molecules in a metal matrix by metal vapor deposition. This method has not yet been perfected. Abbott considered freeze-drying the most promising approach, and he has obtained field-electron-emission molecular images after attempting to deposit soluble ribonucleic acid (sRNA), light meromyosin (LMM), or Cu–phthalocyanine molecules on a field-evaporated W tip.

12.4.5 Problems

There are problems of several types associated with the possible application of FIM to the study of biological molecules. Solutions to the particular problem of molecule deposition on the tip have been indicated, as described above. After deposition it would seem that two possible routes are available, i.e., direct imaging of molecules or imaging the structure of a surrounding matrix. The two cases will be considered separately.

Major obstacles to obtaining meaningful direct field-ion images of large organic molecules occur because of the high fields required. Thus, mechanical distortion may be expected and has been reported[38] to occur in other experiments at lower fields. Electrical distortions due to non-uniform charge distribution on the surface of a complicated nonmetallic material may also be a factor, and the threat of field-induced dissociation is real. Field-induced chemical reactions[18] are also possible if the imaging is done by a chemically active gas.

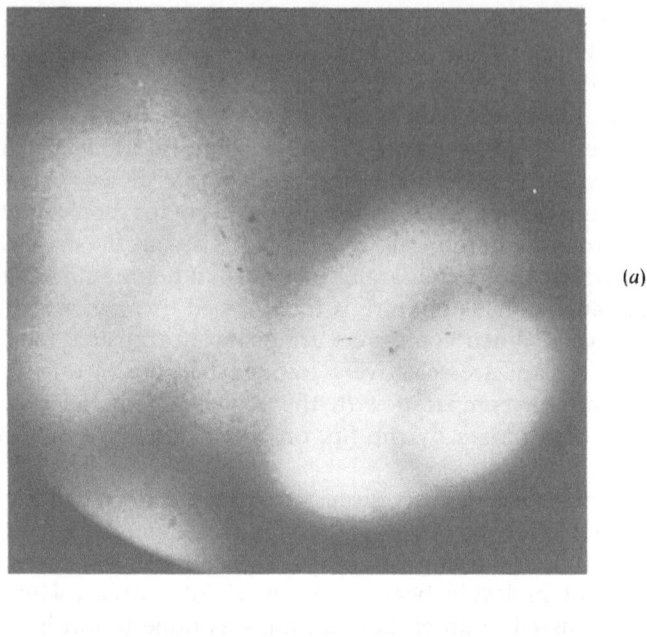

(a)

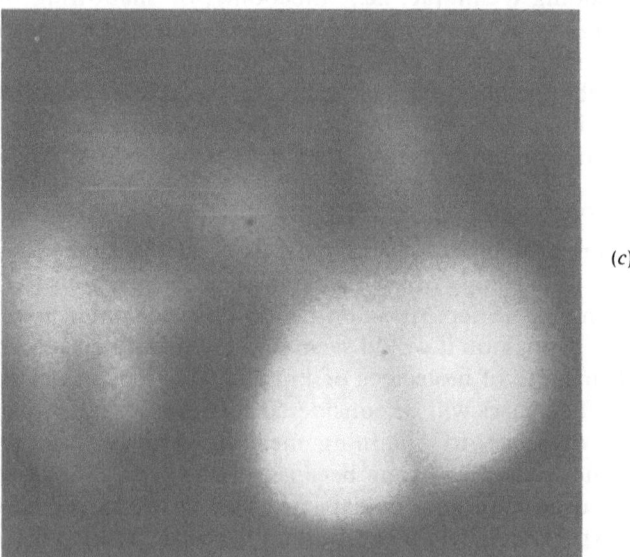

(c)

Fig. 12.18. Nylon adsorbed on W at 77°K: (a, b, c) field-electron micrographs taken about 30 sec apart (unstable image, and (d) hydrogen-ion micrograph (stable image).

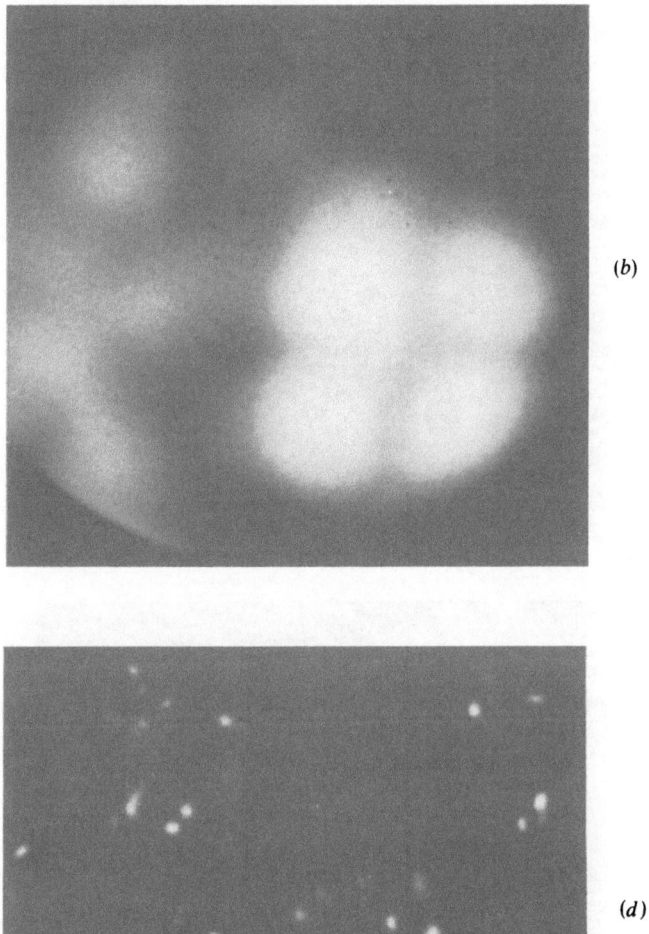

(b)

(d)

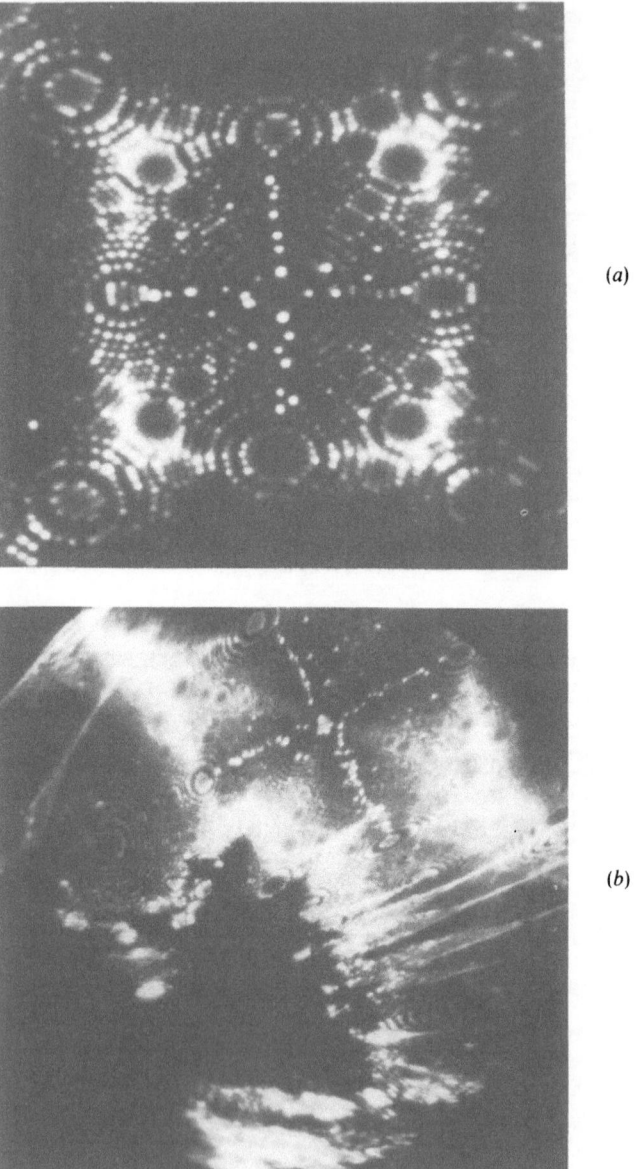

(a)

(b)

Fig. 12.19. Helium field-ion micrographs (77°K) for Ir: (a) field-evaporated surface with high degree of perfection, and (b) large amount of imperfection, including prominent hole, resulting from heating specimen in air prior to resharpening and field evaporation.

If an embedding or shadowing technique is used, the first question mark arises regarding the ability of the surrounding matrix to replicate the molecular shape faithfully. Secondly, interpretation difficulties may be expected in analyzing the matrix images. The micrograph of Fig. 12.19*b* is presented to illustrate part of this problem. It is a helium field-ion micrograph of Ir at 77°K. After tip fabrication and initial FIM imaging (Fig. 12.19*a*), this specimen was heated to about 1500°C in air and then further electrolytically etched and imaged. A large hole, clearly seen in the micrograph, persisted through about 100 atomic layers before the tip was torn off. The edge of the hole is not well defined, and this may be due to the original shape of the hole or it may be an artifact due to enhanced field evaporation at an originally sharp edge. This problem should also exist for the "hole" represented by an embedded molecule. Local differences in field evaporation rates due to imperfections in the surrounding matrix may also be a problem.

In summary, the problem is formidable in view of the present knowledge in field-ion microscopy. However, since the mountain is there, men will try to climb it.

References

1. F. R. N. Nabarro and P. J. Jackson, in: *Growth and Perfection of Crystals*, R. H. Doremus, B. W. Roberts, and D. Turnbull, eds., Wiley (New York), 1958.
2. G. W. Sears, *Acta Met.* **3**: 367 (1955).
3. R. Gomer, *J. Chem. Phys.* **28**: 457 (1958).
4. W. Dittmar and K. Neumann, *Z. Elektrochem.* **64**: 297 (1960).
5. E. W. Müller and O. Nishikawa, *Field Ion Microscopy of Iron Whiskers*, Technical Report AFML-TR-64-346, Air Force Materials Laboratory, Wright-Patterson AFB, Ohio, 1964.
6. A. J. Melmed. *J. Chem. Phys.* **38**: 607 (1963).
7. A. J. Melmed, *J. Appl. Phys.* **34**: 3325 (1963).
8. E. W. Müller, S. Nakamura, O. Nishikawa, and S. B. McLane, *J. Appl. Phys.* **36**: 2496 (1965).
9. A. J. Melmed, *J. Chem. Phys.* **38**: 1444 (1963); *11th Field-Emission Symposium*, Cambridge, England, 1964; *J. Appl. Phys.* **36**: 3585 (1965); *J. Chem. Phys.* **42**: 3332 (1965); *J. Less-Common Metals* **8**: 320 (1965); and A. J. Melmed and R. F. McCarthy, *J. Chem. Phys.* **42**: 1466 (1965).
10. E. W. Müller, *Z. Elektrochem.* **61**: 43 (1957).
11. T. Gurney, Jr., F. Hutchinson, and R. D. Young, *J. Chem. Phys.* **42**: 3939 (1965).
12. T. Rhodin, H. Montague-Pollock, and M. Southon, *12th Field-Emission Symposium*, Pennsylvania State University, University Park, Pa., 1965.
13. R. D. Young and D. C. Schubert, *J. Chem. Phys.* **42**: 3943 (1965).
14. M. Inghram and R. Gomer, *J. Chem. Phys.* **22**: 1279 (1954).
15. R. Gomer and M. Inghram, *J. Amer. Chem. Soc.* **77**: 500 (1955).
16. E. W. Müller and K. H. Bahadur, *Phys. Rev.* **102**: 624 (1956).
17. H. D. Beckey, *Z. Instrumentenk.* **71**: 51 (1963).
18. E. W. Müller, *Advan. Electron. Electron Phys.* **13**: 83 (1960).
19. H. D. Beckey, *Z. Naturforsch.* **17a**: 1103 (1962).
20. R. Gomer, *Field Emission and Field Ionization*, Harvard University Press (Cambridge, Mass.), 1961.
21. H. D. Beckey, *Advan. Mass Spectrometry* **2**: 1 (1963).
22. E. W. Müller, *Naturwisse.* **37**: 333 (1950).

23. J. M. Robertson, *J. Chem. Soc.*, Part I: 615 (1935).
24. R. Haefer, *Acta Phys. Austriaca* **8**: 105 (1953).
25. E. W. Müller, *Ergeb. Exakt. Naturw.* **27**: 290 (1953).
26. P. Wolf, *Z. Angew. Phys.* **6**: 529 (1954).
27. E. Hörl and F. Stangler, *Acta Phys. Austriaca* **10**: 1 (1956).
28. J. A. Becker and R. G. Brandes, *J. Appl. Phys.* **27**: 221 (1956).
29. A. J. Melmed and E. W. Müller, *J. Chem. Phys.* **29**: 1037 (1958).
30. A. J. Melmed, *Study of Phthalocyanine and Some Other Planar Molecules in the Field Emission Microscope*, Technical Report AFOSR TN 58-646, July 1958, ASTIA AD NO: 162 178.
31. A. P. Komar and A. A. Komar, *Zh. Tekhn. Fiz.* **31**: 231 (1961), for English Transl., *Soviet Phys., Tech Phys.*
32. E. W. Müller, *Z. Naturforsch.* **5a**: 475 (1950).
33. E. W. Müller, *Z. Physik* **108**: 668 (1938); also **120**: 270 (1953).
34. D. J. Rose, *J. Appl. Phys.* **27**: 215 (1956).
35. R. C. Abbott and W. A. Livingston, Jr., *12th Field-Emission Symposium*, Pennsylvania State University, University Park, Pa., 1965.
36. E. C. Cooper and E. W. Müller, *Rev. Sci. Instr.* **29**: 309 (1958).
37. A. J. Melmed and E. W. Müller, unpublished results.
38. H. Montague-Pollock, *11th Field-Emission Symposium*, Cambridge, England, 1964.
39. R. C. Abbott, *Rev. Sci. Instr.* **36**: 1233 (1965).

Appendix A

1951–1965

THE FIRST FIFTEEN YEARS OF FIELD-ION MICROSCOPY— A BIBLIOGRAPHY

A.1 General Reviews

1. E. W. Müller, "Das Auflosungsvermogen des Feldionenmikroskopes" ("The Resolution of Field-Ion Microscopes"), *Z. Naturforsch.* **11a**: 88 (1956).
2. E. W. Müller, "Study of Atomic Structure of Metal Surfaces in the Field Ion Microscope," *J. Appl. Phys.* **28**: 1 (1957).
3. E. W. Müller, "Experimenteren mit Atomaren Kristallbausteinen in Feldionenmikroskop," *Z. Electrochem.* **61**: 43 (1957).
4. M. Drechsler, "Kristall Stufen von I bis 1000 Å," *Z. Electrochem.* **61**: 48 (1957).
5. J. A. Becker, "Study of Surfaces by Using New Tools," *Solid State Phys.* **7**: 416 (1958).
6. E. W. Müller, W. T. Pimbley, and J. F. Mulson, "The Study of Metal Surfaces by the Field-Ion Microscope," in: *Internal Stress and Fatigue in Metals*, G. M. Rassweiler and W. L. Grube, eds., Elsevier (Amsterdam), 1959, p. 189.
7. E. W. Müller, "Field Ionization and Field Ion Microscopy," *Advan. in Electron. and Electron Phys.* **13**: 83 (1960).
8. R. Gomer, *Field Emission and Field Ionization*, Harvard University Press (Cambridge, Mass.), 1961.
9. E. Sugata and S. Nakamura, "Study of Field-Emission Cathode by Field-Ion Microscopy," *Appl. Phys. Japan (Oyo Butsuri)* **1**: 50 (1962).
10. D. G. Brandon, "The Resolution of Atomic Structure: Recent Advances in Theory and Development of the Field Ion-Microscope," *Brit. J. Appl. Phys.* **14**: 474 (1963).

11. B. Ralph and D. G. Branson, "The Field Ion Microscope: 1. Design and Development, 2. Applications," *J. Roy. Microscop. Soc.* **82**: 179 and 188 (1964).
12. E. W. Müller, *Progress and Problems in Field Ion Microscopy*, Xerox–Cornell Material Science Center, Lecture Series, Rept. No. 276.
13. E. W. Müller, "Field Ion Microscopy," *Science* **149**: 591 (1965).
14. S. S. Brenner, "Field Ion Microscope Studies of Surfaces," in: *Surfaces: Structure, Energetics, and Kinetics*, Oct. 27, 1965.

A.2 Ion Mechanics: Field Ionization and Field Evaporation

15. E. W. Müller and K. Bahadur, "Velocity Distribution in Field Ion Emission," *Phys. Rev.* **99**: 1651 (1955).
16. E. W. Müller, "Resolution of the Atomic Structure of a Metal Surface by the Field-Ion Microscope," *J. Appl. Phys.* **27**: 474 (1956).
17. E. W. Müller and K. Bahadur, "Field Ionization of Gases at a Metal Surface and the Resolution of the Field-Ion Microscope," *Phys. Rev.* **102**: 624 (1956).
18. E. W. Müller, "Field Desorption," *Phys. Rev.* **102**: 618 (1956).
19. E. W. Müller and J. F. Mulson, "Surface Structure of Field-Evaporated Metal Crystals," *Bull. Am. Phys. Soc., Ser.* 11, **3**: 69 (1958).
20. E. W. Müller, "Perfection of Metal Crystal Surfaces by Field Evaporation," *Bull. Am. Phys. Soc., Ser.* 11, **4**: 322 (1959).
21. E. W. Müller and R. D. Young, "Determination of Field Strength for Field Evaporation and Ionization in the Field-Ion Microscope," *J. Appl. Phys.* **32**: 2425 (1961).
22. T. C. Clements and E. W. Müller, "Occurrence of H_3^+ in the Field Ionization of Hydrogen," *J. Chem. Phys.* **37**: 2684 (1962).
23. M. J. Southon and D. G. Brandon, "Current Voltage Characteristics of the Helium Field-Ion Microscope," *Phil. Mag.* **8**: 579 (1963).
24. G. Erhlich and F. G. Hudda, "Promoted Field Desorption and the Visibility of Adsorbed Atoms in the Ion Microscope," *Phil. Mag.* **8**: 1587 (1963).
25. R. Gomer and L. W. Swanson, "Theory of Field Desorption," *J. Chem. Phys.* **38**: 1613 (1963).
26. E. W. Müller, "The Effect of Polarization, Field Stress and Gas Impact on the Topography of Field Evaporated Surfaces," *Surface Science* **2**: 484 (1964).
27. E. W. Müller, S. Nakamura, and O. Nishikawa, "Field-Evaporation End Form of Pure Metals," *Bull. Am. Phys. Soc., Ser.* 11 **9**: 150 (1964).
28. T. T. Tsong and E. W. Müller, "Measurement of Energy Distribution in Field Ionization," *J. Chem. Phys.* **41**: 3279 (1964).
29. D. G. Brandon, "The Structure of Field Evaporated Surfaces," *Surface Science* **3**: 1 (1965).

A.3 Image Interpretation

30. E. W. Müller, "Extreme Stress Conditions at the Tip Crystal of the Field-Ion Microscope," *Bull. Am. Phys. Soc., Ser.* 11, **3**: 265 (1958).
31. M. Drechsler and P. Wolf, "Zur Analyse von Feldionenmikroskop—Aufnahmen mit Atomaren Auflosung," in: *Intern. Conf. Electron Microscopy, 4th Berlin, Germany*, 1958.
32. A. J. W. Moore, "The Structure of Atomically Smooth Spherical Surfaces," *Phys. Chem. Solids* **23**: 907 (1962).
33. D. G. Brandon, "Image Formation in the Field-Ion Microscope," *Phil. Mag.* **7**: 1003 (1962).
34. D. G. Brandon, "The Accurate Determination of Crystal Orientation from Field-Ion Micrographs," *J. Sci. Inst.* **41**: 373 (1964).
35. S. Ranganathan, "Contrast from Imperfections in Field-Ion Microscopy," in: *Electron Microscopy, 1964, Proc. European Conf., 3rd, Prague, Czechoslovakia*, Publishing House of the Czechoslovak Academy of Sciences, 1964, p. 265.
36. B. Ralph, "The Interpretation of Field-Ion Microscope Images of Alloys," in: *Electron Microscopy, 1964, Proc. European Conf., 3rd, Prague, Czechoslovakia*, Publishing House of the Czechoslovak Academy of Sciences, 1964, p. 265.

37. D. G. Brandon, "The Analysis of Field Evaporation Data from Field-Ion Microscope Experiments," *Brit. J. Appl. Phys.* **16**: 683 (1965).
38. S. Ranganathan, K. M. Bowkett, J. Hren, and B. Ralph, "The Interpretation of Field-Ion Micrographs: Streak Contrast," *Phil. Mag.* **12**: 841 (1965).

A.4 Techniques

39. E. W. Müller, "Das Feldionenmikroskop," *Z. Physik.* **131**: 136 (1951).
40. E. W. Müller, "Betriebsbedingungen des Tieftemperatur—Feldionenmikroskop," *Ann. Physik.* **20**: 315 (1957).
41. E. C. Cooper and E. W. Müller, "Field Desorption by Alternating Fields," *Rev. Sci. Instr.* **29**: 309 (1958).
42. B. J. Waclawski and E. W. Müller, "Operation of the Field-Ion Microscope with a Dynamic Gas Supply," *J. Appl. Phys.* **32**: 1472 (1961).
43. D. G. Brandon, S. Ranganathan and D. S. Whitmell, "Image Intensification in the Field-Ion Microscope," *Brit. J. Appl. Phys.* **15**, 55 (1964).
44. O. Nishikawa and E. W. Müller, "Operation of the Field-Ion Microscope with Neon," *J. Appl. Phys.* **35**, 2806 (1964).
45. C. Baker and B. Ralph, "A Combined Electron and Field-Ion Microscopic Study of Graphite Whiskers," in: *Electron Microscopy, 1964, Proc. European Conf., 3rd, Prague, Czechoslovakia*, Publishing House of the Czechoslovak Academy of Sciences, 1964, p. 325.
46. S. B. McLane, E. W. Müller, and O. Nishikawa, "Field-Ion Microscopy with an External Image Intensifier," *Rev. Sci. Instr.* **35**: 1297 (1964).
47. W. T. Pimbley and R. M. Ball, "Use of a Refrigerator with the Field-Ion Microscope," *Rev. Sci. Instr.* **36**: 225 (1965).
48. B. Ralph and M. J. Southon, "Field-Ion Microscope," *J. Sci. Instr.* **42**: 543 (1965).
49. E. W. Müller and O. Nishikawa, "Increased Image Brightness in a Field-Ion Microscope," *Rev. Sci. Instr.* **36**: 556 (1965).
50. H. F. Ryan and J. Suiter, "An All Metal Field-Ion Microscope," *J. Sci. Instr.* **42**: 645 (1965).
51. V. G. Weizer, "Variable Image Intensification in the Field-Ion Microscope," *J. Appl. Phys.* **36**: 2090 (1965).
52. E. W. Müller, S. Nakamura, O. Nishikawa, and S. B. McLane, "Gas Surface Interaction and Field-Ion Microscopy of Non-refractory Metals," *J. Appl. Phys.* **36**: 2496 (1965).

A.5 Lattice Imperfections

53. M. Drechsler, G. Pankow, R. Vanselow, "Uber den Nachweis von Versetzungen beim Abbau von Wolfram-, Tantal- und Nickel-Einkristallen" ("Concerning the Appearance of Dislocations after Field Evaporation with W, Ta, and Ni"), *Z. Physik. Chem. (Frankfurt)* **4**: 17 (1955).
54. E. W. Müller, "Pseudospirals, Imperfect Structures and Crystal Habit Produced by Field Evaporation of Metal Crystals," *Acta Met.* **6**: 620 (1958).
55. E. W. Müller, "Beobachtungen der Atomartig Struktur von Metalloberflachen im Feldionenmikroskop," *Proc. Intern. Conf. Electron Microscopy, 4th, Berlin, Germany, 1958*, Vol. 1; Springer Verlag (Berlin), 1960, p. 820.
56. E. W. Müller, "Beobachtung von nahezu fehlerfreien Metallkristallen und von Punkt-defekten im Feldionenmikroskop" ("Observation of Nearly Perfect Metal Crystals and of Point Defects in the Field-Ion Microscope"), *Z. Physik* **156**: 399 (1959).
57. E. W. Müller, "Field-Ion Microscope Studies of Surface Corrosion, of Interstitials, Vacancies, and α-irradiation Damage by Controlled Field Evaporation of Atomic Layers," in: *Structure and Properties of Thin Films*, C. A. Neugebauer, J. D. Newkirk, and D. A. Vermileya, eds. Wiley (New York), 1959, p. 476.
58. D. G. Brandon and M. Wald, "The Direct Observation of Lattice Defects by Field-Ion Microscopy," *Phil. Mag.* **6**, 1035 (1961).
59. E. W. Müller, "Direct Observation of Crystal Imperfection by Field-Ion Microscopy," in: *Imperfection in Crystals*, J. B. Newkirk and J. H. Wernick, eds., Wiley (Interscience) (New York), 1961.

60. D. G. Brandon, M. Wald, B. Ralph, and M. J. Southon, "The Application of Field-Ion Microscopy to Some Metallurgical Problems," in: *Proc. Intern. Congr. Electron Microscopy, 5th*, 1962, pp. J–17.
61. E. W. Müller, "Field Ion Microscopy of the Defect Structure of Metal Crystals," *J. Phys. Soc. Japan* **18** Sup. II: (1963).
62. E. W. Müller, "Field Emission Microscopy of Clean Surface with Electrons and Positive Ions," *Ann. N.Y. Acad. Sci.* **101**: 585 (1963).
63. D. G. Brandon, M. Wald, M. J. Southon, and B. Ralph, "The Application of Field-Ion Microscopy to the Study of Lattice Defects," *J. Phys. Soc. Japan* **18**, Sup. II: 324 (1963).
64. E. W. Müller, "Field-Stress-Induced Surface Defects," *Bull. Am. Phys. Soc., Ser.* 11, **9**: 104 (1964).
65. E. W. Müller, "Field-Ion Microscopy of Rhenium," in: *Electron Microscopy, 1964, Proc. European Conf., 3rd, Prague, Czechoslovakia*, Publishing House of the Czechoslovak Academy of Sciences, 1964, p. 161.
66. H. F. Ryan and J. Suiter, "Field-Ion Microscope Observations of Stacking Faults in Tungsten," *J. Less-Common Metals* **9**: 258 (1965).
67. H. F. Ryan and J. Suiter, "Cavities in Tungsten," *J. Less-Common Metals* **9**: 307 (1965).
68. S. Nakamura and E. W. Müller, "Field Evaporation and Form of Tantalum," *J. Appl. Phys.* **36**: 2535 (1965).

A.6 Grain Boundaries

69. T. H. George, "An Unusual Example of a Grain Boundary," *Z. Physik* **176**: 556 (1963).
70. D. G. Brandon, B. Ralph, S. Ranganathan, and M. Wald, "A Field-Ion Microscope Study of Atomic Configuration at Grain Boundaries," *Acta Met.* **12**: 813 (1964).
71. H. F. Ryan and J. Suiter, "Grain Boundary Topography in Tungsten," *Phil. Mag.* **10**: 727 (1964).
72. S. Ranganathan and A. H. Cottrell, "A Field-Ion Microscopic Study of Grain Boundaries in Iridium," in: *Electron Microscopy, 1964, Proc. European Conf., 3rd, Prague, Czechoslovakia*, Publishing House of the Czechoslovak Academy of Sciences, 1964, p. 163.
73. J. Hren, "An Analysis of the Atomic Configuration of an Incoherent Twin Boundary with the F.I.M.," *Acta Met.* **13**: 479 (1965).

A.7 Alloys

74. B. Ralph and D. G. Brandon, "A Field-Ion Microscope Study of Some Tungsten Rhenium Alloys," *Phil. Mag.* **8**: 919 (1963).
75. B. Ralph and D. G. Brandon, "A Field-Ion Microscope Study of the Order–Disorder Reaction in Equiatomic Cobalt–Platinum," *Journées Internationales des Applications du Cobalt* **9**: 1 (1964).
76. B. Ralph and D. G. Brandon, "A Field-Ion Microscopic Study of the Equiatomic Cobalt–Platinum Alloy in the Permanent Magnetic State," in: *Electron Microscopy, 1964, Proc. European Conf., 3rd, Prague, Czechoslovakia*, Publishing House of the Czechoslovak Academy of Sciences, 1964, p. 303.
77. E. K. Caspary and E. Krautz, "Feldionenmikroskopische Untersuchungen im Mischkristallsystem Wolfram Molybdan," *Z. Naturforsch.* **19a**: 591 (1964).

A.8 Radiation Damage

78. E. W. Müller, "Observation of Radiation Damage with the Field Ion Microscope," in: *Reactivity of Solids, Proc. Intern. Symp. Reactivity of Solids, 4th*, J. H. de Boer *et al.*, eds., Elsevier (Amsterdam), 1960, p. 691.
79. D. G. Brandon, M. J. Southon, and M. Wald, "The Application of Field-Ion Microscopy to Radiation Damage," in: *Proc. Intern. Conf. Berkeley Castle, Gloucestershire, England*, Butterworth (London), 1961, p. 113.
80. M. K. Sinha and E. W. Müller, "Bombardment of Tungsten with 20 keV Helium Atoms in a Field-Ion Microscope," *J. Appl. Phys.* **35**: 1256 (1964).

81. K. M. Bowkett, J. Hren, and B. Ralph, "A Study of Neutron Damage with the Field-Ion Microscope," in: *Electron Microscopy*, 1964, *Proc. European Conf., 3rd, Prague, Czechoslovakia*, Publishing House of the Czechoslovak Academy of Sciences, 1964, p. 191.
82. M. J. Attardo and J. M. Galligan, "Radiation Damage in Platinum," *Phys. Rev. Letters* **14**: 641 (1965).
83. K. M. Bowkett, L. T. Chadderton, H. Norden, and B. Ralph, "A Study of Fission Fragment Damage in Tungsten with the Field-Ion Microscope," *Phil. Mag.* **11**: 651 (1965).

A.9 Other Applications

Adsorption, Corrosion, Surface Diffusion, Whiskers

84. E. W. Müller, "Observation of Paired Screw Dislocations in Iron Whiskers," *J. Appl. Phys.* **30**: 1843 (1959).
85. G. Ehrlich and F. G. Hudda, "Observation of Adsorption on an Atomic Scale," *J. Chem. Phys.* **33**: 1253 (1960).
86. M. Drechsler, "Uber Versetzeugen in Eisen-Whiskern nach Feldionenmikroskop—Aufnahmen," *Phys. Verhandlungen* **3**: 115 (1962).
87. G. Ehrlich and F. G. Hudda, "Direct Observation of Individual Adatoms: Nitrogen on Tungsten," *J. Chem. Phys.* **36**: 3233 (1962).
88. H. D. Beckey, "Field Ionization Mass Spectroscopy," in: *Advan. Mass Spectrometry* **2**: Pergamon Press (New York), 1962.
89. H. D. Beckey, "Feldionisations–Massenspektren organischer Molekule. I. *n*-Paraffine von C_1 bis C_9" ("Field Ion Mass Spectroscopy of Organic Molecules I. *n*-Paraffins from C_1 to C_9"), *Z. Naturforsch.* **179**: 1103 (1962).
90. H. D. Beckey and G. Wagner, "Analytische Anwendungsmoglichkeiten des Feldionen–Massenspektrometers" ("Possibilities for Analytical Applications of the Field Ion Mass Spectrometer"), *Z. Anal. Chem.* **197**: 58 (1963).
91. D. W. Bassett, Thermal Rearrangement of a Perfectly Ordered Tungsten Surface," *Nature* **198**: 468 (1963).
92. A. J. Melmed, "Field Electron and Field-Ion Emission from Single Vapour-Grown Whiskers," *J. Chem. Phys.* **38**, 607 (1963).
93. A. J. Melmed, "Field-Emission Microscopy of Twins in Vapour-Grown F.C.C. Whiskers,"
94. J. F. Mulson and E. W. Müller, "Corrosion of Tungsten and Iridium by Field Desorption of Nitrogen and Carbon Monoxide," *J. Chem. Phys.* **38**: 2615 (1963).
95. H. D. Beckey, "Production of the Ionized State of Molecules by High Electric Fields," *Bull. Soc. Chim. Belges.* **73**: 326 (1964).
96. G. Ehrlich, "An Atomic View of Adsorption," *Brit. J. Appl. Phys.* **15**: 349 (1964).
97. D. W. Bassett, "The Thermal Stability and Rearrangement of Field Evaporated Tungsten Surfaces," *Proc. Roy. Soc. (London)* **A256**: 191 (1965).
98. H. D. Beckey and G. Wagner, "Feldionen–Massenspektren organischer Molekule. II. Amine" ("Field Ion Mass Spectroscopy of Organic Molecules. II. Amines"), *Z. Naturforsch.* **20a**: 169 (1965).
99. H. D. Beckey, "Analyse fester organischer Naturstoffe mit dem Feldionen–Massenspektrometer" ("Analysis of Solid Organic Substances with the Field Ion Mass Spectrometer"), *Z. Anal. Chem.* **207**: 99 (1965).
100. H. D. Beckey, "Fieldionen–Massenspektren organischer Molekule. III. *n*-Paraffine bis zum C_{16} und verzweigte Paraffine," *Z. Naturforsch.* **20**: 1329 (1965).
101. H. D. Beckey and P. Schulze, "Feldionen–Massenspektren organischer Molekule. IV. Olefin," *Z. Naturforsch.* **20a**: 1335 (1965).
102. S. Nakomura and E. W. Müller, "Initial Oxidation of Tantalum Observation in a Field-Ion Microscope," *J. Appl. Phys.* **36**: 3634 (1965).
103. T. Gurney, F. Hutchinson, and R. D. Young, "Condensation of Tungsten on Tungsten in Atomic Detail: Observation with the Field-Ion Microscope," *J. Chem. Phys.* **42**: 3939 (1965).
104. R. D. Young and D. C. Schubert, "Condensation of Tungsten on Tungsten in Atomic Detail: Monte Carlo and Statistical Calculation *vs* Experiment, *J. Chem. Phys.* **42**: 3943 (1965).

A.10 Author Index

Appendix B

LATTICE GEOMETRY

B.1 Plane Spacings

The value of d, the distance between adjacent planes in the set (hkl), may be found from the following equations:
Cubic:

$$\frac{1}{d^2} = \frac{h^2 + k^2 + l^2}{a^2}$$

Hexagonal:

$$\frac{1}{d^2} = \frac{4}{3}\left(\frac{h^2 + hk + k^2}{a^2}\right) + \frac{l^2}{c^2}$$

B.2 Cell Volumes

The following equations give the volume v of the unit cell:
Cubic:

$$v = a^3$$

Hexagonal:

$$v = \frac{\sqrt{3a^2c}}{2} = 0 - 866a^2c$$

B.3 Interplanar Angles

The angle ϕ between the plane $(h_1k_1l_1)$ and the plane $(h_2k_2l_2)$ may be found from the following equations:
Cubic:

$$\cos\phi = \frac{h_1h_2 + k_1k_2 + l_1l_2}{\sqrt{(h_1^2 + k_1^2 + l_1^2)(h_2^2 + k_2^2 + l_2^2)}}$$

Hexagonal:

$$\cos\phi = \frac{h_1h_2 + k_1k_2 + \frac{1}{2}(h_1k_2 + h_2k_1) + (3a^2/4c^2)l_1l_2}{\sqrt{[h_1^2 + k_1^2 + h_1k_1 + (3a^2/4c^2)l_1^2][h_2^2 + k_2^2 + h_2k_2 + (3a^2/4c^2)l_2^2]}}$$

Appendix C

**Angles between Crystallographic
Planes in Crystals of the Cubic System**

{HKL}	{hkl}	Values of angles between HKL and hkl planes (or directions)					
100	100	0°	90°				
	110	45°	90°				
	111	54° 44′					
	210	26° 34′	63° 26′	90°			
	211	35° 16′	65° 54′				
	221	48° 11′	70° 32′				
	310	18° 26′	71° 34′	90°			
	311	25° 14′	72° 27′				
	320	33° 41′	56° 19′	90°			
	321	36° 43′	57° 42′	74° 30′			
110	110	0°	60°	90°			
	111	35° 16′	90°				
	210	18° 26′	50° 46′	71° 34′			
	211	30°	54° 44′	73° 13′	90°		
	221	19° 28′	45°	76° 22′	90°		
	310	26° 34′	47° 52′	63° 26′	77° 5′		
	311	31° 29′	64° 46′	90°			
	320	11° 19′	53° 58′	66° 54′	78° 41′		
	321	19° 6′	40° 54′	55° 28′	67° 48′	79° 6′	
111	111	0°	70° 32′				
	210	39° 14′	75° 2′				
	211	19° 28′	61° 52′	90°			
	221	15° 48′	54° 44′	78° 54′			
	310	43° 5′	68° 35′				
	311	29° 30′	58° 31′	79° 58′			
	320	61° 17′	71° 19′				
	321	22° 12′	51° 53′	72° 1′	90°		
210	210	0°	36° 52′	53° 8′	66° 25′	78° 28′	90°
	211	24° 6′	43° 5′	56° 47′	79° 29′	90°	
	221	26° 34′	41° 49′	53° 24′	63° 26′	72° 39′	90°
	310	8° 8′	58° 3′	45°	64° 54′	73° 34′	
	311	19° 17′	47° 36′	66° 8′	82° 15′		
	320	7° 7′	29° 45′	41° 55′	60° 15′	68° 9′	75° 38′ 82° 53′
	321	17° 1′	33° 13′	53° 18′	61° 26′	70° 13′	83° 8′ 90°

Angles between Crystallographic
Planes in Crystals of the Cubic System (*Continued*)

{HKL}	{hkl}	Values of angles between HKL and hkl planes (or directions)						
211	211	0°	33° 33′	48° 11′	60°	70° 32′	80° 24′	
	221	17° 43′	35° 16′	47° 7′	65° 54′	74° 12′	82° 12′	
	310	25° 21′	49° 48′	58° 55′	75° 2′	82° 35′		
	311	19° 8′	42° 24′	60° 30′	75° 45′	90°		
	320	25° 9′	37° 37′	55° 33′	63° 5′	83° 30′		
	321	10° 54′	29° 12′	40° 12′	49° 6′	56° 56′		
		70° 54′	77° 24′	83° 44′	90°			
221	221	0°	27° 16′	38° 57′	63° 37′	83° 37′	90°	
	310	32° 31′	42° 27′	58° 12′	65° 4′	83° 57′		
	311	25° 14′	45° 17′	59° 50′	72° 27′	84° 14′		
	320	22° 24′	42° 18′	49° 40′	68° 18′	79° 21′	84° 42′	
	321	11° 29′	27° 1′	36° 42′	57° 41′	63° 33′	74° 30′	
		79° 44′	84° 53′					
310	310	0°	25° 51′	36° 52′	53° 8′	72° 33′	84° 16′	
	311	17° 33′	40° 17′	55° 6′	67° 35′	79° 1′	90°	
	320	15° 15′	37° 52′	52° 8′	74° 45′	84° 58′		
	321	21° 37′	32° 19′	40° 29′	47° 28′	53° 44′	59° 32′	
		65°	75° 19′	85° 9′	90°			
311	311	0°	35° 6′	50° 29′	62° 58′	84° 47′		
	320	23° 6′	41° 11′	54° 10′	65° 17′	75° 28′	85° 12′	
	321	14° 46′	36° 19′	49° 52′	61° 5′	71° 12′	80° 44′	
320	320	0°	22° 37′	46° 11′	62° 31′	67° 23′	72° 5′	90°
	321	15° 30′	27° 11′	35° 23′	48° 9′	53° 37′	58° 45′	63° 36′
		72° 45′	77° 9′	85° 45′	90°			
321	321	0°	21° 47′	31°	38° 13′	44° 25′	50°	60°
		64° 37′	69° 4′	73° 24′	81° 47′	85° 54′		

Appendix D

STANDARD STEREOGRAPHIC PROJECTIONS

The field-ion image approximates a stereographic projection. Since the wires that are used for the preparation of field-ion specimens usually exhibit strong textures, the standard projections that prove particularly useful are (100), (111), (110) for cubic crystals and (11$\bar{2}$0) for hexagonal crystals. Since hexagonal crystals specially grown to have (0001) as the wire axis are available, this projection is also included. It may be noted that the planes that are prominent in field ion images from f.c.c. crystals are marked in the (100) and (111) projection, while the (110) projection has planes that are prominent in images from b.c.c. crystals.

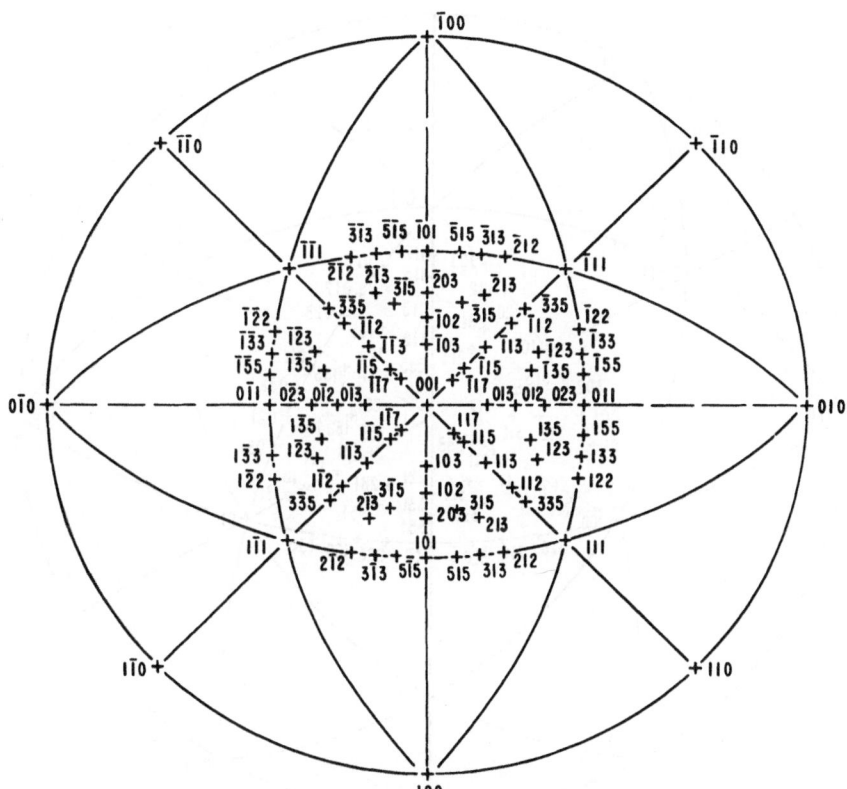

Fig. D.1. Standard (001) projection for cubic crystals.

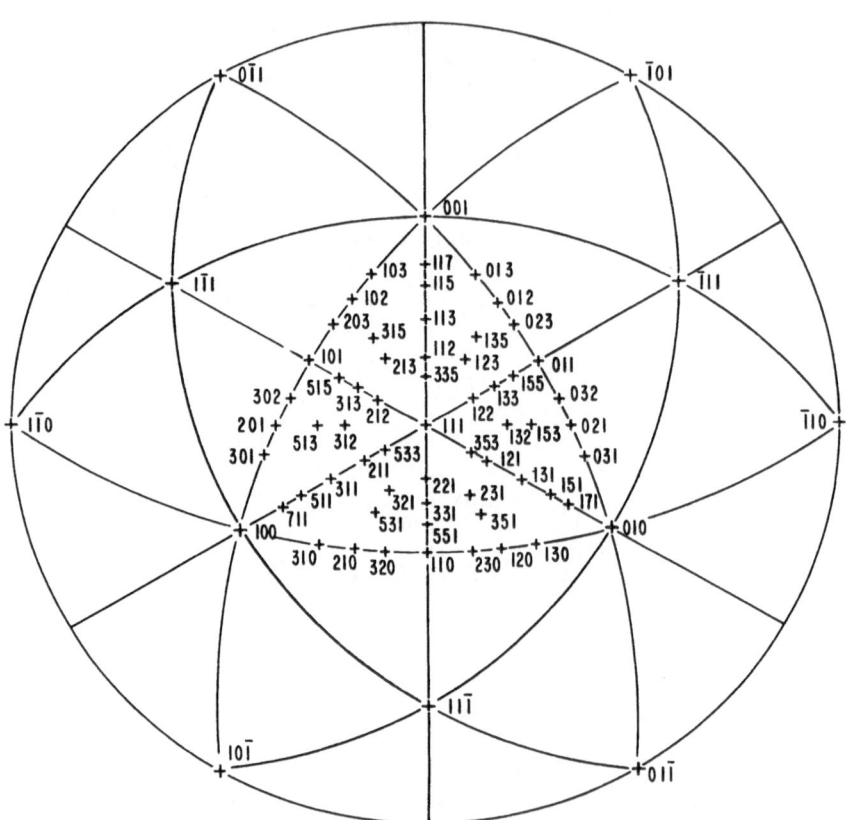

Fig. D.2. Standard (111) projection for cubic crystals.

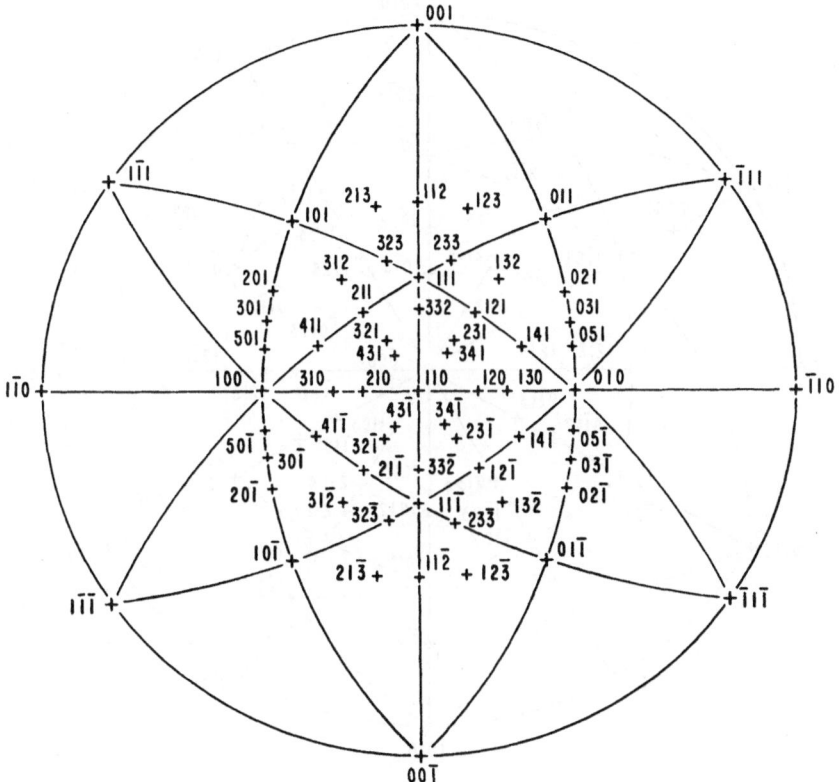

Fig. D.3. Standard (110) projection for cubic crystals.

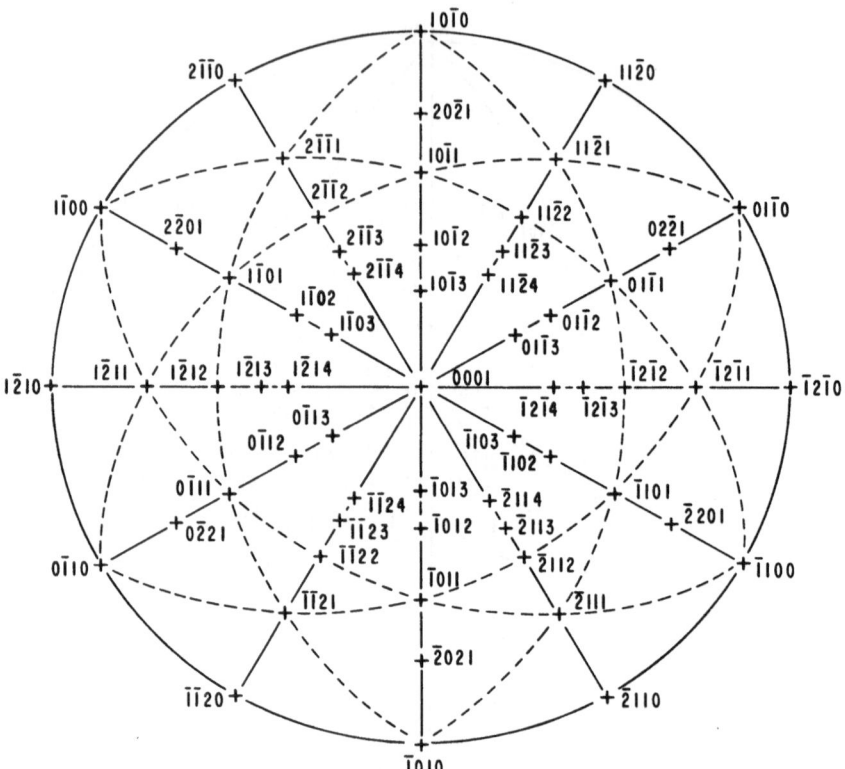

Fig. D.4. Standard (0001) projection for hexagonal crystals. (The ratio c/a is the ideal one for hexagonal close packing.)

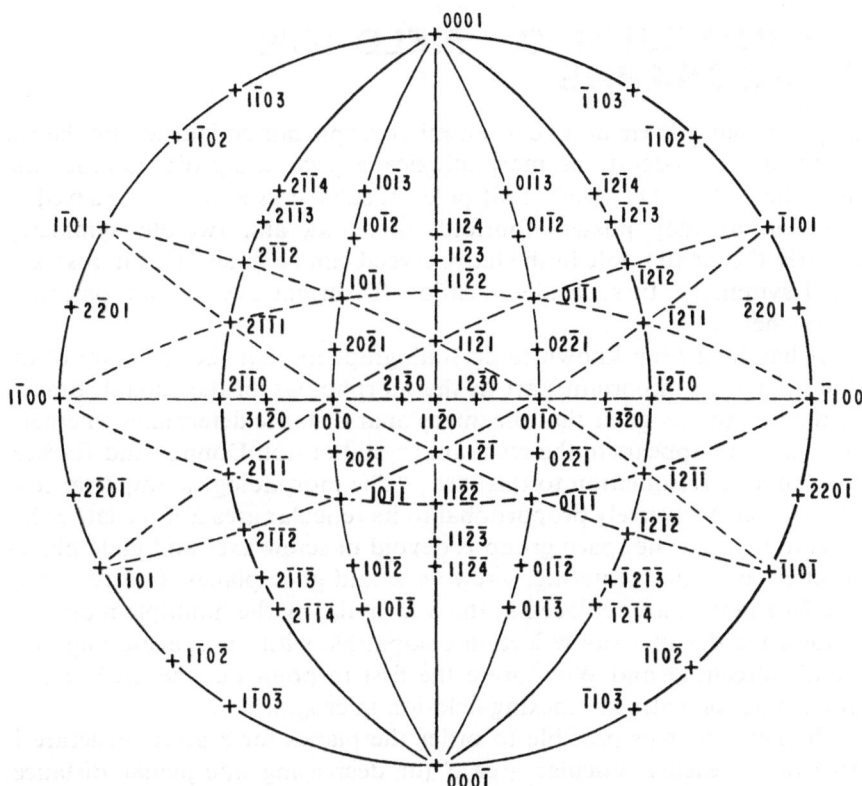

Fig. D.5. Standard (11$\bar{2}$0) projection for hexagonal crystals. (The ratio c/a is the ideal one for hexagonal close packing.)

Appendix E

THE INDEXING OF FIELD-ION MICROGRAPHS

The poles that appear in a field-ion micrograph can be indexed by the use of symmetry considerations, morphological aspects and projection relationships. The {100}, {111}, and {110} poles in cubic crystals can be indexed by inspection as they possess fourfold, threefold, and twofold symmetry. Similarly the {0001} pole in the h.c.p. crystal can be indexed, as it possesses sixfold symmetry. In such identification the bright *zone-decoration* atoms are very helpful.

It has long been known to crystallographers that the structure of the crystal plays an important part in the morphology of the crystal. Bravais was the first to recognize that the interplanar distance determines whether a particular face appears in the resultant crystal or not. Donnay and Harker[1] gave a precise formulation to this idea: "The morphological importance of a crystal face is inversely proportional to its reticular area S if the lattice has no centering and the space group is devoid of screw axes and glide planes. The effect of lattice centering, screw axes, and glide planes is corrected for if the face indices are replaced in the S formula, by the 'multiple indices' of the lowest order of x-ray reflection compatible with the space-group symmetry." Drechsler and Wolf[2] were the first to point out the usefulness of x-ray extinction rules in indexing field-ion micrographs.

It then becomes possible to order the planes for a given structure in terms of increasing reticular area S (or decreasing interplanar distance). Such a list is also an indication of the morphological importance of various planes. The lists for b.c.c. and f.c.c. crystals are given below.

Body-Centered Cubic

hkl	110	200	211	310	222	321	411	420	332	431
S^2	2	4	6	10	12	14	18	20	22	26

Face-Centered Cubic

hkl	111	200	220	311	331	420	422	511	531	442
S^2	3	4	8	11	19	20	24	27	35	36

Thus it is possible to distinguish between b.c.c. and f.c.c. field-ion micrographs by merely noting whether a twofold or a threefold axis is important. An indication of the importance of a particular plane in the micrograph is given by the number of rings having the same pole.

The morphological aspect can thus be used for indexing a micrograph. However the projection relationships are also useful in such an indexing process. This is briefly considered below.

The field-emission microscope presents a less complicated geometry than the field-ion microscope. The screen is in the shape of a hemisphere, and the photographed image is an orthographic projection of the crystal being imaged. The field-ion microscope uses a flat screen, as the low image intensity necessitates the use of f : 1 lenses with a shallow depth of focus. Müller[3] made the assumption that the field-ion micrograph is also a case of orthographic projection. Brenner[4] has shown that the projection is nearer the stereographic projection by the simple expedient of superposing a stereographic net of the appropriate size on the field-ion image. Recently

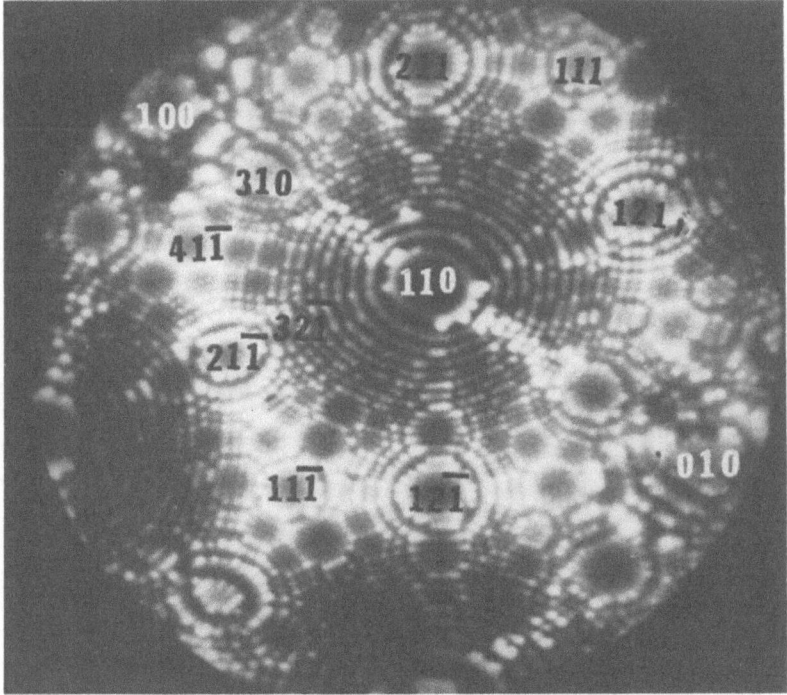

Fig. E.1. Helium field ion micrograph of tungsten with prominent planes indexed.

Brandon[5] has shown that the distances of poles from the central pole are best explained by assuming that the projection center is two radii away from the center. (For the three classical projections—the gnomonic, the stereographic, and the orthographic projection—the projection center is zero, unit radius, and infinite distance from the tip center). The result was entirely empirical. Unless great accuracy is demanded, the field-ion micrograph can be treated as a case of stereographic projection. Figure E.1 shows a tungsten field-ion micrograph indexed on the above principles.

References

1. J. D. H. Donnay and D. Harker, *Am. Mineralogist* **22**: 446 (1937).
2. M. Drechsler and P. Wolf, *Proc. Intern. Conf. Electron Microscopy, 4th, Berlin*, Springer Verlag (Berlin), 1958, p. 835.
3. E. W. Müller, *Advan. Electron. Electron Phys.* **13**: 83 (1960).
4. S. Brenner, *Metal Surfaces*, ASM publication, 1962.
5. D. G. Brandon, *J. Sci. Instr.* **41**: 373 (1964).

Appendix F

Polishing Solutions and Conditions

Material	Electrolyte	Remarks
Tungsten	5% NaOH	5–15 V d.c.
Tungsten	20% KCN	1–5 V a.c.; start at higher voltage
W–Re alloys	20% KCN	As for tungsten
Tantalum and its alloys	90% HNO_3 10% HF or 17.5% HF 17.5% H_2SO_4 65% H_2O	0–3 V d.c.; very low current densi- ties chilled electrolyte in a stain- less steel beaker
Niobium	Molten $NaNO_2$	6 V a.c.
Niobium and its alloys	As for tantalum and its alloys	
Rhenium	Conc. HNO_3	10 V d.c.
Iridium	20% KCN	3–15 V a.c.; start at higher voltage
Molybdenum	20% KCN	1–5 V a.c.; start at higher voltage
Platinum	Molten NaCl	5.5–6 V d.c.
Platinum	20% KCN	3–15 V a.c.; start at higher voltage
Pt–Co alloys	20% KCN	3–15 V a.c.; start at higher voltage
Zirconium	10% HF	Dip into solution
Beryllium	Conc. H_3PO_4	30–50 V d.c.
Rhodium	Aqu. sol. KCN	1 V a.c.
Silicon	45 pts 40% HF 60 pts Conc. HNO_3 20 pts acetic acid 3 pts bromine	Dip into fresh sol.
Gold	50% HCl, 50% HNO_3	10 V a.c.
Gold	20% KCN	3–10 V a.c.; start at higher voltage
Iron	10% HCl	1–3 V a.c.; start at higher voltage
Cobalt	10% HCl	4–6 V d.c.
Titanium	40% HF	4–12 V d.c.
Palladium	30% HCl, 70% HNO_3	3 V a.c.
Nickel	10% HCl	1–3 V a.c.; start at higher voltage
Copper	Conc. H_3PO_4	1–5 V a.c.
Zinc	Conc. KOH	10–15 V d.c.
Tin	40% HF	1–6 V a.c.

Appendix G

MICROSCOPE DESIGNS

There are at least as many field-ion microscope designs as there are researchers. References 39 through 52 of Appendix A are an excellent source for particular design features. Commercial microscopes are also available from the following:

CENCO Instruments, 1700 Irving Park Road, Chicago, Illinois, 60613 (U.S.A.)

HRB–Singer, Box 60, State College, Pennsylvania 16801 (U.S.A.)

Jackson and Church Electronics, 1127 South Patrick Drive, Satellite Beach, Florida 32935 (U.S.A.)

Materials Research Corporation, Route 303, Orangeburg, New York 10962 (U.S.A.)

Optometric Instruments, 8255 Beverly Boulevard, Los Angeles, California 90048 (U.S.A.)

Twentieth Century Electronics Ltd., King Henry's Drive, New Addington, Croydon, Surrey (U.K.)

The following figures are intended to illustrate the range of designs of varying complexity that are appropriate for particular applications.

Fig. G.1. A bakeable glass field-ion microscope with liquid-nitrogen cooling. (Courtesy of A. J. W. Moore.)

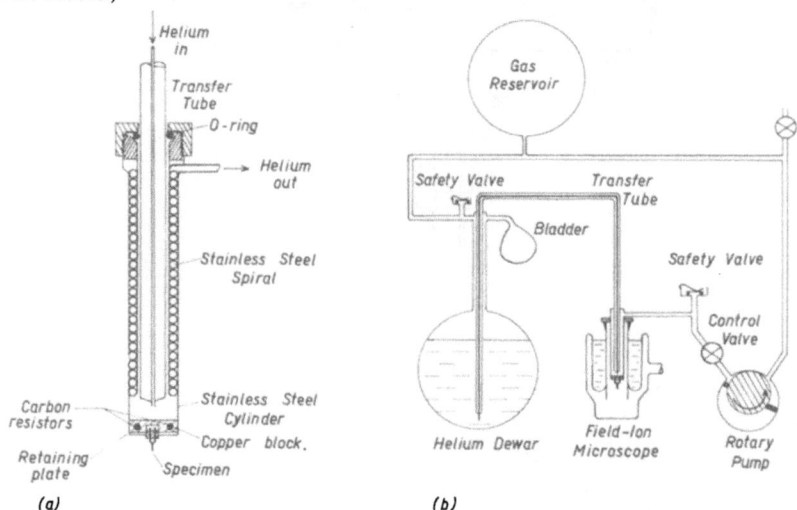

Fig. G.2. Schematic for a liquid-helium-cooled field-ion microscope of glass. (Courtesy of D. G. Brandon.)

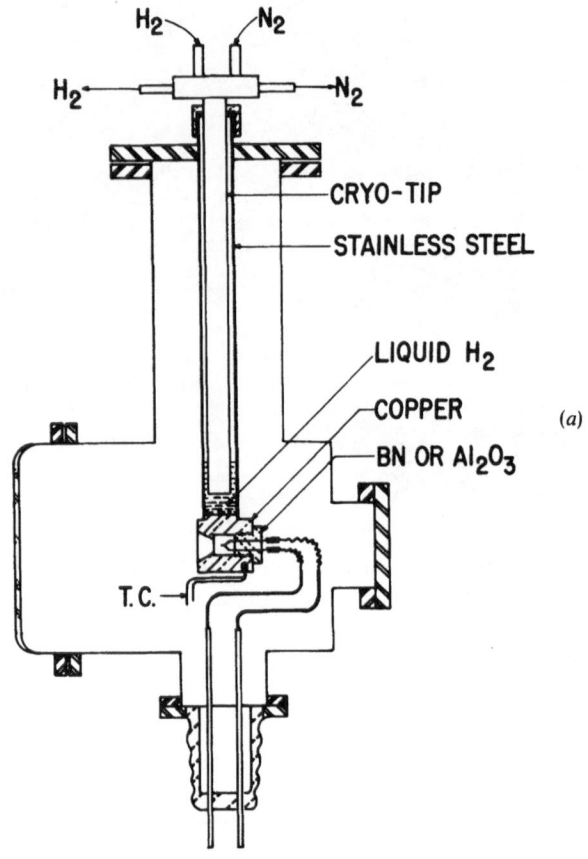

(a)

Fig. G.3. Liquid-hydrogen-cooled stainless steel field-ion microscope: (a) schematic of microscope body, and (b) photograph of the system. (Courtesy of S. S. Brenner.)

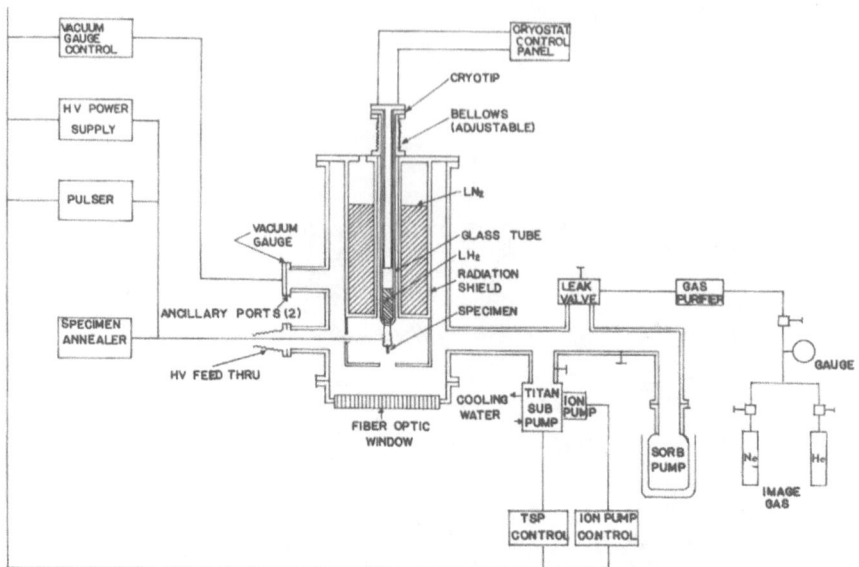

Fig. G.4. Functional diagram of combined field-ion and field-electron microscope with UHV system, voltage pulser, and specimen heater. (Courtesy of Jackson and Church Electronics.)

INDEX